CW01221380

Leatherhead Food International

FUNDAMENTALS OF FOOD REACTION TECHNOLOGY

Mary Earle and
Richard Earle

This edition first published 2003 by
Leatherhead Publishing
a division of
Leatherhead International Limited
Randalls Road, Leatherhead, Surrey KT22 7RY, UK
URL: http://www.lfra.co.uk

and

Royal Society of Chemistry,
Thomas Graham House, Science Park, Milton Road, Cambridge CB4 0WF, UK
URL: http://www.rsc.org
Registered Charity No. 207890

ISBN No: 1 904007 53 8

A catalogue record of this book is available from the British Library

Typeset by Leatherhead International Ltd.
Printed and bound in the UK by Antony Rowe Ltd., Bumper's Farm, Chippenham, Wiltshire, SN14 6LH.

CONTENTS

PREFACE

Industrial food processing has been a craft and an art, but is rapidly moving towards being a modern technology. One clear way to meet the enhanced sensory quality, safety, nutrition, health, economy, and novelty demanded of food products by consumers is by improving the operations of food processing. Improvement is related to the better prediction and control that follow quantitative description of the reactions, the chemistry changes that occur during the processing of food materials – the rates of changes and the factors influencing the changes. This is the fundamental reason for writing this book at the present time, to indicate the wealth of knowledge becoming available on reactions in food processing, and the use of reaction technology to apply this knowledge in food processing. It is important that not only operations managers and technologists, but also development technologists and engineers, consider how reaction technology impacts on their present and future processes and products.

This book introduces the methods of reaction technology, setting its essence out simply and concisely, and illustrating how it has been and can be applied in industrial practice. It builds a framework for the application of reaction technology, and then uses this in straightforward, hopefully readily understandable examples with an industrial context. The numerical detail seeks to bring out a sense of magnitudes and rates appropriate to real systems, and to develop a feel for required processing precision matched both to the product quality outcomes and also to the necessities of the processing. It tries also to lead the reader to consider ways in which the examples can be extended into wider applications, with the intention that they will in turn point the way to more day-to-day use in food processing operations. In some areas of food processing, reaction technology has been used for many years, because it has provided the reliability and safety that consumers demanded. Commercial sterilisation in canning was a very early pioneer; it has been joined by shelf-life prediction, and process control in the dairy industry. But these are largely isolated from the vast bulk of food processing operations, and even from each other by using different, and therefore potentially confusing, nomenclatures. There is a need to make reaction technology in food processing a coherent whole.

Many years ago after writing a small introductory book on unit operations in food processing, it was gratifying to see that it filled an apparent and substantial need and was very widely used across the world. Not only was it found applicable to the academic context for which it was intended, but it also found acceptance in a food industry that was becoming much more aware of how a structured and systematic technology could help in solving processing problems. At the time, a parallel need could be discerned for looking into reactions of change in food constituents that were undergoing processing in the multitude of vessels, packages, and containers that are the reactors of the food industry. As with the unit operations, one of the great benefits of such an approach is to unify and thereby strengthen and reinforce the theory. It is then a short step to applications through borrowings and adaptations that assert themselves as obvious. Using such an approach in teaching and research over the years has further reinforced this. It demonstrated that students could appreciate and apply reaction technology directly to food processing. It was also continually amplified by the emergent literature. This literature has, after hesitant beginnings, come almost universally to work with and quote the standard methodology of chemical reaction engineering because that methodology is so direct and so applicable to the manifest needs of food processing.

It has always been a problem that, because of the innate complexity of so many food systems, full technological analysis can rapidly become too complicated for industrial application. This means that, to apply reaction technology in food processing, there must be careful selection of the important reactions, recognition of the levels of process accuracy necessary to achieve the specified product qualities, and use of modern information technology to analyse and control the process. Obviously, skill and knowledge are needed to design and operate a process using reaction technology, and this book introduces the knowledge and some of the skills in application required. It is not a textbook on reaction technology in food processing; that is left for the future.

The book sets out the general principles governing changes in the nature, the chemistry, of a food constituent and then extends this to include the dynamics of the reactions of the many chemical constituents of food raw materials and ingredients. It does this quantitatively because that is what process technology is all about. It demonstrates that many other important attributes, including microbiological safety and consumer acceptability, can very often be fitted comfortably into the general framework. It is illustrated by references from the publications of researchers, many of them chosen from the recent literature. Where it seemed that there might be interest in background detail of theory, this has been touched on but in detached sections that can be skipped by those who are willing to accept results on trust.

Chapter 1 introduces the broad concepts relating the particular food situation to the general framework of reaction technology. There is more complexity than in most chemical technology, because of the composite nature of the raw material reactants. Adding this to the biological origins and the instability of food raw materials increases complications but creates no fundamental differences that cannot be incorporated or blended into the analysis. Included are not only typical food processing situations, but also extensions to new product and process design, shelf life and storage, process control, product quality and safety.

Chapter 2 shows how component concentrations in the foods are fitted to the rate equations, and basic concepts such as reaction order and activation energy are introduced. The rate equations lead to prediction of changes in concentrations of components, rates of changes and how these rates can be varied under the control of the industrial processor. The rates of change, their sensitivities to external influences such as the temperature and the times over which the process operates, determine the relative constitution and thus the properties of the final food product. Practical industrial food situations are used as illustrations.

Chapter 3 considers integration of the rate equations to predict the product outcomes. Analytical, numerical and graphical methods are described, and then applied to design and to control practical food processing situations. Because temperature conditions are often not uniform throughout the food, such as in a can or a loaf of bread, impacts of the differences on the reaction rates can be important. Important raw material variables extend beyond chemical concentrations, so the application of reaction technology to microbiological growth and death is also analysed.

Chapter 4 then extends the range from single reactions to multiple reactions, studying a number of reactants in series or in parallel, and to the relative rates of these. A number of reactions, selected as the critical ones for the product attributes and quality, are studied simultaneously and optimum processing conditions determined. Attributes such as the level of viable pathogenic or spoilage microorganisms, nutritional value, sensory assessments such as colour and flavour, and even general acceptability, can be included and assigned priorities to arrive at improved products. Examples of these are introduced and discussed.

Chapter 5 extends reaction technology from heat processing to other methods of processing using processing agents, and alternative energy sources such as irradiation, electrical and magnetic fields, and very high pressures. These are fitted into the general reaction technology framework. They can be used alone or in combinations, and can offer particular advantages since the total processing outcome is the sum of the separate changes, each fitting particular circumstances and contributing their own special features, which can all be analysed.

The book ends with an analysis of several applications of the reaction technology approach, and then demonstrates something of the scope, the adaptability, and the success of the methods of reaction technology in food processing.

The book is in stages, so that those who are interested in the application of reaction technology can take it as far as they need. Those who need more detail for process development will hopefully be encouraged to take further steps. This can lead them into the massive body of published research results, and also the internal resources of data and knowledge that exist in many firms, but which have not been explored to their full capacity. The worked examples have been selected to illustrate applications as well as methods, and they make up an essential part of the text. Processing is practical and based on experimentation and measurement. Mathematical analysis of the results can give the basis for prediction of effects on product quality of changes in process variables, and therefore the basis for both process design and control. Calculations can be quick and painless on modern hand calculators, or by using widely available computer software such as spreadsheets and graphics. From the predictions can come better and more consistent products, more efficient processes, and ideas that can contribute to the development of new products and processes. When approached systematically, it is perhaps surprising and certainly stimulating to see how much order can be applied through reaction technology to the apparently diverse food processing and preservation activities ranging across industries, institutions, restaurants, and homes.

The book is intended for industrial technologists working in process design, organisation and control. It is only an introduction, an appetiser. But it may serve to stimulate hunger to pick up and to make fuller use of what is understood and has been uncovered, and to give some direction towards what new knowledge needs to be found and developed for the future processing of better foods. It is hoped that it will also give technical managers an overall view of how the application of reaction technology in the future can lead to a "high tech" food industry.

Acknowledgement

Inevitably the contents of this book are built on the work of others. The names of all of them are far too numerous to mention, but some are singled out by inclusion in the references and examples. It is hoped that they have not been misrepresented or misquoted. That the literature exists is a huge resource for industry, and hopefully books such as this, suggesting the possibility of further uses for the published results, will help to make it even more useful. A great debt is owed to all who have worked and contributed to bring this knowledge to its present level.

The presentation used in this book was developed in undergraduate teaching in food and bio-processing in New Zealand and in Canada. Postgraduate students and industrial consultations in New Zealand and Britain enhanced it. It was crystallised by presentation in seminars to university teachers in Thailand, who found it useful in building their courses in food processing. Those participating have contributed to and enhanced the material, and for this we are grateful.

1. IMPORTANT PROBLEMS IN FOOD PROCESSING

1.1 Introduction

Food processing includes all the activities that control the nature of food between the agricultural and marine production and its final eating by the consumers. It includes everything from the controlled conditions in the transport and storage of whole fresh meat, fish, fruit and vegetables, to the complex processing producing food ingredients followed by manufacturing to produce the final consumer product. Before being eaten, biological materials from agriculture or fishing are transformed through processing into the finished foods the consumer wants. Food processing makes the food products more attractive, more satisfying, safer and easier to eat, and preserves them from deterioration. It includes building up desirable constituents and removing or reducing undesirable ones, encouraging enzymes to develop desirable flavours and textures and removing or inhibiting enzymes causing undesirable changes, growing microorganisms to create flavour and texture and destroying them to prevent harm to the consumer or decay of the food.

Food products are the outcomes of food processing, and it is important to identify the desirable product qualities and the undesirable and even unsafe product qualities. The products are the aim of food processing, and processing needs to be designed and controlled to give the product qualities identified and wanted by the consumers. Food processing is diverse, complex, and often carried out on a large industrial scale.

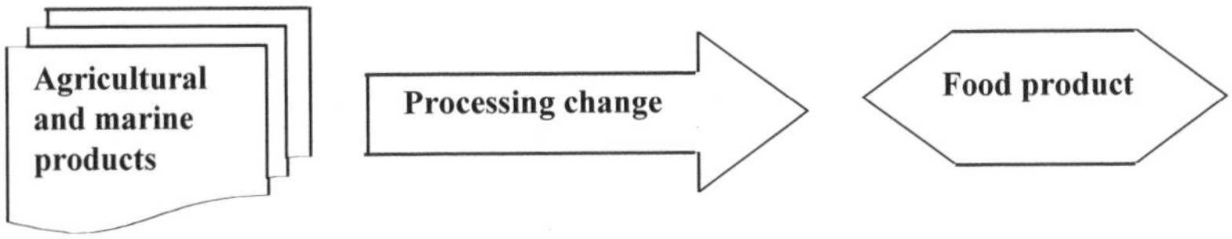

Figure 1.1. Conversion in food processing

1.2 Changes During Food Processing

Processing causes changes in the food materials; some of the changes are shown in Table 1.I.

TABLE 1.I
Changes in food materials during processing

Chemical:	hydrolysis, oxidation, polymerisation, denaturation, de-amination, browning, hydrogenation, esterification, destruction of toxic substances.
Physical:	gelation, hardening, softening, toughening, emulsifying, colour loss/gain.
Biological:	growth and death of microorganisms, glycolysis, physiological changes in ripening.
Nutritional:	constituent availability, protein changes, loss of vitamins, amino acids loss, destruction of anti-nutritional substances.
Sensory:	aroma and flavour loss, aroma and flavour changes, texture changes, colour bleaching and darkening.

From Earle & Earle (1)

These changes can be measured, so their progress during processing can be followed and studied by the food technologist. The progress of processing can be measured in many ways, such as chemical analysis, physical measurements, counts of microorganisms, and colour, texture and flavour assessments by sensory panels. Changes can often be described in terms of the changing chemical composition, that is changes in the concentrations of the chemical components, but sometimes this is not possible and sensory, physical or microbiological measurements are used to quantify the changes.

Measurement reveals continuing change with time during the process. As our knowledge extends over ever-wider ranges of foods and food processing, and our analytical skills increase, the measured changes are increasingly found to be systematic and describable in quantitative terms. The quantitative data from change measurement can be fitted to mathematical equations and to physical models. The models can be tested and, if necessary, modified until they fit observations adequately for practical processing purposes. Once the models are sufficiently established, they can be used to predict changes in processing between and sometimes beyond the original processing conditions. The models can be employed industrially to guide the processing, to control its extent, and to design new processes and equipment. They can predict outcomes under different processing conditions, conditions that can be set before the processing is started and regulated until completion. Important processing variables include temperature, time, moisture level, pH and atmosphere, and different levels can be set to give the processing conditions.

The changes start when the process begins, and move on through the processing towards defined ends. The changes differ in their desirability between what are called customarily food processing and food preservation. Food processing, as seen traditionally, is about causing wanted changes in the food as it moves towards a finished product. These changes improve the food, adding to its value. For economic reasons, they often need to happen speedily. Food preservation aims to slow down undesirable changes, and conditions have to be organised so that the changes happen as little as possible. They are deleterious to the quality and value of the food. The changes are normally spontaneous, arising from the instability of the food, and the processing conditions are arranged to slow them down. Since they are both about change, its manipulation and its control, and since they can both be technologically described in the same way, it is convenient to think of the whole area as that of dynamic food processing.

In all dynamic food processing, the aim is always towards a defined product outcome. In food processing, the defined end is an optimum food product. In food preservation, the defined end is a point at which the food becomes significantly less edible or desirable, reaching a minimum acceptable quality. This is the point at which quality is measurably degraded and which the preservation process is designed to avoid reaching. Both involve changes that the technologist seeks to understand and keep under control. The changes take place under the scientific laws that govern reactions. They are influenced both by material qualities and by processing conditions, many of which are, or can be brought under the control of the processor. This book looks at the ways in which these changes occur, at quantitative descriptions of them in simple terms that can be used in practice, and at examples of industrial application.

Think break

Select two food-processing operations with which you are familiar.

* Identify all the changes that take place in the raw materials as they are moved through processing towards finished foods.
* Identify the individual chemical constituents so far as you can and the changes in these that occur.
* Consider the ways in which the changes are regulated and under control from the beginning to the end point in each stage of the process.

As an introduction, a number of important challenges involving the dynamics of change in food processing have been selected and will be outlined. Specific examples illustrate these challenges, focusing on food products and on food processing. They will show how, in particular industrial situations, the challenges have been studied using the methods of process reaction technology. In the later chapters there is more detailed discussion of how these methods can be applied generally and specifically to a range of food processes.

1.3 Food Products

Food products cover all edible products in the food system: industrial, foodservice and consumer products, primary produce, food ingredients, retail foods, and domestic foods. At the end, consumers determine the qualities expected of these products, but intermediate customers in the food system, such as food processors, food manufacturers and food retailers, very often set the working product specifications. Although these products may differ a great deal, their basic qualities can be grouped into composition, nutritional value, sensory, safety and health. In studying food processing, it is important firstly to identify the specific critical and important food qualities, called *product attributes* (or characteristics), required in each product, then to set the optimum values of the product attributes, so that food processing can be designed and controlled to attain the specified values for these specific attributes.

1.3.1 Consumer expectations

Consumer product expectations are built up from eating, or from publicity in the case of new and untried products. Consumer concerns include nutrition, food safety, shelf-life, as well as social and environmental aspects (2), but customers also very much want desirable sensory qualities and psychological benefits in the food. Expectations are becoming more specific all the time as consumers' knowledge of food qualities increases. They want specific qualities and they want them to be true for all units of the products. For example, in the supermarkets, the customer wants today's product to be true to the type and the quality bought last week or last month, and to a high degree of precision unless good reasons are produced for any change and such changes are acceptable.

Consumer expectations translate directly into buying specifications. When these product quality demands are combined with formal consumer requirements to list compositions and nutritional contents on packets, the food manufacturer has to pay attention, often minutely, to processing changes and to bring them under very precise control. The only real way to do this efficiently is to seek detail of, and manipulate, the process conditions as carefully as is currently possible.

1.3.2 Product attributes

A food product can be viewed at three levels – the consumer's product concept, the total company product, and the company's basic functional product (2). In setting the product aims for food processing, the consumer's product expectations have to be converted into the company's basic functional product, which is described by its physical, chemical, microbiological, sensory and nutritional attributes, as shown in Fig. 1.2.

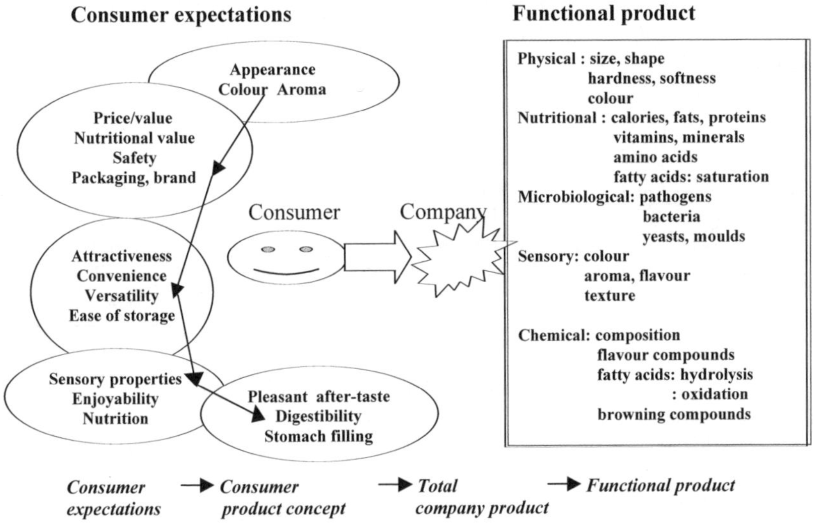

Fig. 1.2. Building the functional product from consumer expectations

General properties such as liking, use, convenience, safety, health, storage life, and consistency of quality and safety have to be converted into quantitative measures of specific product attributes with a required statistical framework. Sensory characteristics, such as appearance, aroma, flavour and texture, are developed into physical measurements, chemical constituents or sensory scores. The critical and important attributes, such as protein content, hardness, specific or general bacterial levels, acidity, solubility, and significant flavour(s), have to be identified for each product, and the required levels of these built into the final product specifications.

In studying food processing, it is important to identify:

- the critical and the important attributes of the final product
- levels of the critical and important attributes that are acceptable
- sensitivity of the product attributes to changes in processing conditions.

1.3.3 Product specifications

The industrial food processor combines ingredient product attributes into product buying specifications; today, these are very often set in conjunction with their industrial customers. The food manufacturer defines the product specifications for the final consumer products using the consumer knowledge from product development, but also with regard to regulatory and retailer requirements. Such groups as supermarket chains and multinational food manufacturers routinely use sophisticated product specifications. To produce food products with the required levels of attributes in the product specifications, the reactions producing these attributes need to be understood and controlled in the processing. This is illustrated in Fig. 1.3.

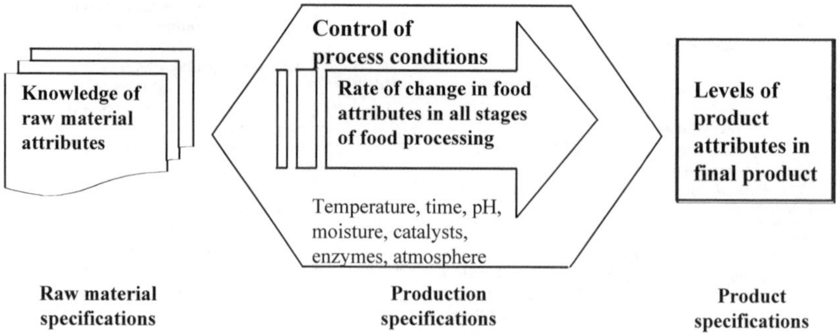

Fig. 1.3. Specifications in food processing

The sheer power of modern chemical analyses going down to parts per billion and the use of such tools as mass spectrometers, tristimulus colour meters, bacterial serotyping, and the continual emergence of new techniques, have provided the wherewithal for the buyer to discriminate almost infinitely. Additionally, if customer-buying specifications were not sufficiently demanding, governments and regulatory authorities are reacting to the substantial pressures imposed on them and exercising their powers ever more extensively. They have justifiably serious concerns for maintaining public health and safety. This is generating fresh stipulations all the time, for example on newly uncovered pathogens and including such processing-resistant entities as the prions, suspected of causing *bovine spongiform encephalopathy* (mad-cow disease) and thence perhaps variant Creutzfeldt-Jakob disease in people eating beef. In addition, regulations are being increasingly demanding on health statements and implications in advertising, for example for nutraceuticals and other foods claiming to be active for this or against that. All of these imply detail in food contents, which have to be properly certified by the processors and legally defended if needed. These requirements are demanded, not only in the food as it is initially placed on the shelves of supermarkets, but also increasingly until some nominated date thereafter.

Each process usually affects several specified attributes, which can be classified as critical, important and unimportant. For example, some of the attributes in orange juice are described in Example 1.1; there are important attributes, such as flavour and aroma, related to consumer acceptability, and a critical attribute, pathogenic organisms that could cause food poisoning.

Example 1.1: Important and critical attributes of orange juice

Fruit juices, such as orange juice, when newly extracted have the flavours and aromas of the original fruit. Many of the especially attractive features diminish gradually with time. The period of time the flavours and the aromas are retained depends on the particular fruit and ambient factors such as temperature and oxygen access. By using techniques such as chromatography, the gradual flavour losses can be monitored, and, by reference to sensory panels, quality thresholds can be set up. For example, the times needed under particular conditions to reach the level of flavour loss that is just detectable to trained panellists can be used as guides and references by the industry and retailers. Also, at the time of extraction, actual and potential off-flavours can arise, such as from citrus oils in the peel of the fruit.

More critically, pathogenic microorganisms can contaminate the juice. These pathogens constitute such a significant health hazard that, in some jurisdictions, notably the USA, legal regulations demand that the packed juice must be subjected to a process sufficient to reduce the numbers of the most resistant pathogens in the finished juice by a factor of five log. cycles, or else the food must be given a warning label that may be detrimental to marketing. This imposes a requirement on the processor to identify the most resistant pathogens in the product, and to ascertain the kinetics of their destruction. Any approved process can be used to ensure that the product meets the stipulations of the regulation with regard to pathogen reduction. This process may be pasteurisation, which is heat treatment, but it may also be one of a number of non-thermal processes, such as high pressure or irradiation.

Adapted from Mermelstein (3)

The attributes measurable in a food can be ranked as critical, where levels are mandatory (because of safety, regulations, contractual stipulations, company policies, strong consumer reaction), very important, where they are significant contributors to quality/value/market appeal, and then range down to unimportant. In processing, critical attribute levels must be maintained, very important ones should be maintained (subject to economic/plant criteria), whereas unimportant ones may not even be recorded (unless used as indicators of processing or product improvement).

Think break

Drink some UHT liquid milk and:

* Identify all the sensory attributes from first looking at the milk in the glass to the final swallowing.

Contd..

Think break (contd)

* Group the sensory attributes into critical, important and unimportant.
* Choose the sensory product attributes to use for measuring the effect of changes in processing conditions.
* Decide what are the critical safety and nutritional attributes that must also be used to control the process.

1.3.4 Sensitivity of product attributes to processing conditions

A continuing problem in processing is assessing the sensitivity of critical and important attributes of the final product to changes in the controllable process variables such as temperature, atmosphere, moisture, catalysts, enzymes and, most importantly, time.

For example, bakers have seen batches of freshly baked bread and biscuits emerging from the oven with the colour of the crusts somewhere between overcooked and burned, and pondered hard whether they should all be rejected, or the darkest culled out, or maybe all of them let go to the market? In any event, it signals losses, even if only in consumer satisfaction. Appearances are important, and the surface on the top of the loaf or the biscuit is just so prominent. Obviously, temperatures have been too high somewhere in the oven or the loaves were spending too long there. Significant questions lie not only in the setting of the temperatures and time but also in the precision of controlling them. The complex relationships between time and temperatures determine the extent of the browning reactions that bring colour and flavour to the crust of the bread, but can also lead quickly to burning and charring. Would a drop in the oven temperatures (and if so how much) and/or shortening the times (and if so how much) allow the oven to produce paler and more consistent crust colour? But would this affect the texture of the loaf, making it doughy? Or would the shelf life of the loaf be reduced by growth of moulds?

These same broad questions permeate much of processing. 'Trial and error' experimentation will always bring solutions, but these experiments may be quite lengthy and expensive to work through, and they may still end up some distance from the best solution. When there are several simultaneous reactions proceeding in some process, there are trade-offs that have to be made amongst the processing conditions and they can almost always be made in a host of ways. Which are better, and is there a best? The sensitivities of the various critical and important attributes to processing conditions such as time and temperature need to be evaluated before selecting the optimum process conditions. Optimum process conditions for one attribute may be less than optimum for another, so compromises need to be made. In Example 1.2, this is discussed in regard to the pasteurisation of milk, one of the early examples of the use of reaction technology in processing, although such implications were only partially appreciated at that time.

Example 1.2: Pasteurisation of milk – choosing the processing conditions

For many years, and in most countries, cows' milk for human consumption has been treated by pasteurisation, in which it is subjected to controlled heating. This is motivated almost totally for health reasons, and has had a demonstrable and beneficial effect on public health. Therefore, the criteria for the treatment were dominated by the conditions that provided the consumer with safety against organisms of concern, such as *Mycobacterium tuberculosis* and *Coxiella burnettii*. Regulations to demand a specific processing condition of temperature and time for liquid milk were imposed, in some countries virtually as soon as knowledge to do this was available.

Experimentation later showed that a number of process conditions could provide equal protection from the microorganisms. Processes stipulating higher temperatures and shorter times became possible, especially with the evolution of plate heat exchangers. Although clinical and public health opinion was conservative, and regulations were not easy to change, the high-temperature/short-time processes were accepted by the regulators.

All of these processes caused sensory changes to the milk quality such as some browning and caramelisation giving a 'cooked' flavour, as well as some nutritional changes and destruction of vitamins and enzymes. To be on the safe side, most processors erred on the side of over-processing. And so milk with appreciable cooked flavours became fairly general and it was often all but impossible to buy anything else.

However, investigation of the various processes occurring during the pasteurisation showed that the browning, vitamin, enzyme and bacterial changes behaved systematically. Therefore, by using reaction technology, the sensitivity of the various reactions to the processing conditions could be found. A range of new process conditions were then specified that guaranteed that the product would agree with the product specifications set by the company, regulators and the consumer. Under these processing conditions of time and temperature, the required health stipulations could be met, but the extent of other detrimental reactions and thus of unwanted changes in sensory and nutritional attributes could at the same time be minimised. With knowledgeable process control design and operation, the sensory changes were virtually imperceptible to the customer. This created a superior product for the consumer, but one that was equally and adequately safe.

Adapted from Lewis & Heppell (4)

1.4 New Product Design

Technology push, while being only part of the whole product development (PD) strategy motivation, can be a very powerful tool in the generation of new products. For this to happen, firstly the processing technology has to exist or be invented, and secondly the technology has to be implemented. Scientific and technological knowledge is the basis for the processing technology, and, in particular, reaction kinetics and reaction technology. Every encouragement is needed to expand the fundamental food processing knowledge base so that it can be available for future product design. This knowledge is needed over the broad landscape of food raw materials and the totality of their possible changes and combinations. It can then be focused and directed to particular product development and commercial opportunities, and the technological knowledge implemented in the industrial situation, in the creation of a product that the market will buy and continue to seek. It may seem straightforward, but the implementation generally proves to be the most difficult and the most expensive part of the PD Process in practice.

Therefore, there is a call for more systematic understanding of the processes and the process details, so that processing knowledge can be adapted and applied in new product design, particularly in developing innovative products. Knowledge of rates of reactions can produce a range of new products, such as shown in Example 1.3, where different heat treatment regimes, together with membrane separation, are used to separate proteins from milk and produce a range of food ingredients.

Example 1.3: Milk protein products

Milk contains a considerable range of proteins, and the individual proteins or groups of proteins have all manner of properties that make them both useful and nutritious as ingredients in other foods. Between the various protein constituents, the properties vary considerably, and with them demand and price. It is therefore of real benefit to be able to process milk so that these proteins can be separated, generally into banded groups with specific properties, such as the caseins. It is worthwhile that such groups of proteins, if possible and economic, be commercially produced and marketed to appropriate food manufacturers for incorporation into their products.

Whey was produced in huge quantities in the dairy industry as a by-product of the manufacture of butter and cheese, and was a major waste problem. Research was then undertaken into the nature of the whey proteins, and into their properties. In particular, the effect of varying the time and temperature of the heat treatment, and also the pH, on the rates of denaturation of the different proteins and their precipitation from the whey were studied. These rates were found to vary with the different

Contd..

Example 1.3 (contd)

proteins and led to systematic methods of separating the proteins in the milk, when combined with membrane technology. Both the separation of the protein groups and the properties of the separated proteins were related to the conditions of the heat treatment. Such processing needed to be treated systematically and controlled carefully so as to produce protein products with specific properties.

Over the years, these separations have become more sophisticated and have led to a wide variety of dairy protein ingredients and a substantial industry.

Adapted from Earle, Earle & Anderson (5); Huffman (6)

In the design of new products, careful thought has to be given to the consumer needs and wants and also to the production requirements. This is especially so for the continuous, large-scale operations that mass production and marketing demand. In the food industry, product design and process development have to be integrated throughout the PD Process. Central to the product design specification and its outcomes are:

- the changes that processing is designed to accomplish,
- the ways in which these changes materialise, and
- the controls needed to ensure that they will, and always will, end up with the product that the consumers said they wanted and would buy (7).

Further, knowledge in more detail of these changes in processing can expose practical possibilities, which can be incorporated in the product design in order to make even more exciting and adventurous foods than might otherwise arise.

By their very nature, new products imply creation and invention. These processing activities, bringing in the novel and the previously uncharted, need the widest and deepest tool bags to use in their explorations and trials. Reaction technology with experimental designs can provide the knowledge base for these new developments in the future. In Example 1.4, it is shown that the study of two reaction rates – destruction of pathogenic organisms and protein denaturation – could lead to a new product – eggs that have been pasteurised in their shells to satisfy consumer demands for increased product safety.

Example 1.4: Developing pasteurised eggs in their shells

The potential presence of pathogenic microorganisms in foods is a problem of great importance to human health. These organisms, for example *Salmonella* spp, can be reduced by heat treatments to numbers that make them unimportant as a health problem. But the egg proteins are also denatured by heat treatments.

Investigations into rates of protein denaturation and of pathogenic organisms' destruction during heat treatment showed that, for some important vegetative pathogens, the effect of increasing the temperature on the rate of destruction was significantly less than that on the protein denaturation. This meant that, at longer processing times at lower temperatures, the destruction of pathogens was much faster than the denaturation of proteins. In practice, this implied that long-duration heat processes at relatively modest temperatures could achieve pasteurisation without undesirable concomitant coagulation of proteins.

It was claimed that this could be applied to treatment of eggs in their shells. By this treatment, a new product, an egg that is still soft but is at the same time safely pasteurised, can be produced and marketed.

Adapted from Hou, Singh, & Muriana (8); Polster (9)

Think break

Consider three important product specification issues, for example labelling, health, and safety.

* Relate these to the demands they impose on the two food processing operations lines that you selected in the first think break on page 3.
* How might the processing need to be modified if these specifications were tightened?
* What further information and process knowledge would be needed to enable this to be implemented?
* How might these be related to any costs and disadvantages that might follow non-compliance?

1.5 Product Shelf Life

The time that a food lasts as an acceptable and safe product in distribution, in storage, in marketing and in the home, has always been important. Short shelf life can limit value and availability and may even generate serious shortages when replacement foods are expensive or unobtainable. It is only recently that it has become common, and often legally required, to specify use-by dates or something equivalent on the foods or their packaging. Such claims raise many immediate problems and some ambiguities. Another important determinant of acceptable shelf life is nutritional and compositional labelling. For example, if the levels of labile nutritional compounds such as vitamins change with time and with storage temperatures, does the label carry the initial concentration when packed, the concentration at the nominal expiry data, or some mean concentration between these two? And whichever is chosen, how does it relate to temperature variations in storage, which are all but inevitable under any normal conditions? And if, as say with vitamin A, a high content may be detrimental to health, how are the consumers to evaluate their intakes?

Difficult as these questions may be for the customer, they can be worse for the retailer or the wholesaler faced with queries and complaints, and even, as society becomes more punctilious and litigious, cause expensive proceedings for the food manufacturers. Food manufacturers may have extensive product testing regimes, and temperature- and atmosphere-controlled storage, but there still remain elements of ambivalence that are almost built into the system. The retailer may attempt to opt out, taking large margins of safety by specifying unduly short storage time availability, but this is expensive, and bothersome in terms of stock rotation, may be bad publicity for the store or the product or the product line, and may cause large product returns from the retailer to the manufacturer. Finally, there is the uncertainty once the food gets into the domestic situation, where the potential consumer may be unsure, variable, and capricious in meeting the demands made for storage on the home shelf or refrigerator before consumption. The benefit of any doubt will seldom go to the manufacturer if the food is below par when eaten. 'Keep refrigerated' is of doubtful assistance as home refrigerators are not noted for the high quality of their temperature controls; nor is space always available in them for longer-term holding, however justifiable.

1.5.1 Studying shelf life

In studying the shelf life of a food, there needs to be knowledge of:

- product attributes significant to acceptable quality
- critical and important product attributes, their acceptable levels and any statistical variations in attribute levels that are allowable
- changes of product attributes with time
- reactions causing these changes, many of which are spontaneous reactions and time-dependent
- rates of these reactions, which may vary for different product attributes
- effects of storage variables on rates of reactions.

It is important in studying shelf life to control the storage variables, for example time and temperature, that affect the rate of deterioration. Storage variables are shown in Table 1.II.

TABLE 1.II
Storage variables affecting the shelf life of foods

Food materials	Environment	Packaging
Microbiological quality:	**Temperature**	**Permeability to:**
Pathogens	**Water activity**	Oxygen
Spoilage bacteria	**Atmosphere:**	Water
Yeasts	Oxygen	Carbon dioxide
Moulds	Carbon dioxide	Ethylene
Composition:	Inert gases	Odours
Moisture	Ethylene	Solvents, oils
Acidity/pH	**Light**	**Light transmission**
Sugar	**Microorganisms**	**Packaging migration**
Salt	**Pests**	**Product/packaging interaction**
Preservatives	**Contaminants**	

Adapted from Singh (10); Ellis (11)

Perhaps the first really extensive study reported on shelf lives of foods was carried out in California in the 1950s on frozen foods (12) This examined the premise, then commonly held, that, once fully frozen, foods in general were substantially stable. The researchers studied frozen fruits, vegetables, meats and fish and, by using large resources of people, sensory panels and time, effectively established that this was not true, as shown in Example 1.5.

Example 1.5: Shelf life of frozen foods

In studying the frozen storage of fruits, vegetables, meats and fish, some chemical and sensory attributes were measured. Samples after cold storage at a constant temperature were compared with samples held at a temperature considered low enough for no change to occur. Sensory difference tests and statistical analysis techniques were used to compare the test and control samples. Shelf-life expiry was the earliest time when a difference between the quality of the test and control samples was statistically detected by a panel trained to assess the 'overall quality' of the food.

Major difficulties arose in carrying out this study, in particular the time scale of years found to be needed, and the sensitivity to temperatures that required much more precise and reliable temperature controls than were normal in cold stores. Coupled with this was the need to train sensory panels to give reproducible judgements over these long periods.

Contd..

Example 1.5 (contd)

The research showed essentially that deterioration occurred systematically at rates dependent on temperature relationships well established in the chemical literature. From these results, the practical storage lives for frozen foods were calculated, which have been updated through the years, as shown in Table 1.III.

Adapted from Van Arsdel *et al.* (12); Singh (10); Erickson & Hung (13)

The results of these time-temperature tolerance studies gave both an impetus and a basis to kinetic understanding of food product behaviour on storage. Notably, it proved that shelf lives of food products conformed generally with what would be expected from reaction technology, and were systematically predictable in terms of the food material and the conditions of its storage. They led to the concept of practical storage life (PSL) on which to base storage and distribution specifications. The PSL of frozen foods can be defined as 'the period of frozen storage after freezing of an initially high-quality product during which the organoleptic quality remains suitable for consumption or for the process intended' (14). Some typical PSLs for frozen foods are given in Table 1.III.

TABLE 1.III
Practical storage life of frozen foods

	Practical storage life (months) at		
	-12 °C	**-18 °C**	**-24 °C**
Peaches, apricots, cherries, raw	4	18	>24
in syrup	3	18	>24
Green beans	4	15	>24
Cut corn	4	15	>24
Peas	6	24	>24
Carrots	10	18	>24
Beef carcasses	8	15	24
Beef minced	6	10	15
Fish lean	4	9	>12
Fish fatty, glazed	3	5	>9

From IIR (4)

Shelf-life testing has led to substantial benefits for:

- consumers, so that they gain protection from codes, for example that of the Association of Food and Drug Officials of the United States (AFDOUS), and from adequate labelling of shelf-life expectations at certain temperatures;

- processors, so that they can reliably act to ensure high quality in their products as delivered to the customer;
- retailers and warehousemen, for stock control and rotation;
- designers, so that they can design storage and distribution equipment to deliver designated shelf lives for their products throughout distribution and marketing.

1.5.2 Extending product shelf life

Preservation has always been a major issue in food production and marketing, but the measures taken to extend it in the past were largely empirical and based on observation and *ad hoc* experimentation. Today, better identification of the critical product attributes and the storage variables determining product life allows for more scientific exploration of their changes and dynamics. This allows for the setting of specifications and the use of modern controls, leading to prevention or at least reduction of deteriorative changes. This degree of knowledge may not always be available or determinable in any particular case, but it can be built up through basic research. The real advantage of improved understanding of deterioration rates lies in the much more accurate predictions of these rates and of the resulting shelf lives of the foods. For example, knowledge of the effects of temperature on reaction rates can quantify the effects of reduced temperatures of storage. Also from this reaction rate/temperature relationship can come more precise evaluation of the benefits of any particular temperature level and the extent of tolerance of temperature fluctuations. Because lower temperatures and closer controls cost more, the product quality benefits of these can be balanced against their additional costs, and optimal conditions found and stipulated for warehousing and for protective packaging. An illustration comes from research to improve the quality of fish after catching, as shown in Example 1.6.

Example 1.6: Deterioration of fish after catching

Observation shows that fish are generally in their best condition immediately after catching, and that their quality changes detrimentally thereafter. Over the years, a number of chemical, biochemical, and physical tests have been evolved to measure these changes, and also there has been extensive use of sensory panels. These studies have shown that the acceptability sensory panel scores diminish systematically after catching, and off-odours and off-flavours and their related chemical constituents increase. The rates of these changes are temperature-dependent.

There is, of course, appreciable experimental variability and therefore need for statistical treatment of the data, but they clearly show that the

Contd..

Example 1.6 (contd)

deterioration of the fish quality, by whatever measure, is systematic and statistically reproducible, and that it follows regular kinetics. It shows that the critical variable is the temperatures at which the fish are held for all the time after catching, and that the dependence of rates of deterioration on temperature is in accordance with the general behaviour of chemical reactions.

An important outcome for the product quality is that decreasing the temperature as soon after catching as possible and also having low temperatures in holding and storage, are beneficial. The results allow calculation of the extent of the benefit that can be expected under a particular temperature regime and matching the value of the benefits of lower temperatures against the increased costs of the processing.

Adapted from IIR (15); Earle (16)

In Table 1.III, the storage life of lean frozen fish is shown as increasing from 4 months at -12 °C to 9 months at -18 °C, and greater than 12 months at -24 °C. Fatty fish has shorter frozen storage life, but it is also increased from 3 to 5 to >9 months at, respectively, -12, -18 and -24 °C.

Think break

Compare the use-by (or best-by) dates on ten different chilled and frozen foods in your local supermarket.

* Describe the processing methods for these products.
* How do the use-by dates relate to the processing method?
* Outline how modifying the processing might extend these dates.
* Find the packaging and the storage temperatures for these products.
* Outline how modifying the packaging and the storage temperatures might extend the dates.

1.6 Storage and Distribution Design

The design of foods includes the design of product, process, and package, as well as the conditions in the distribution system. A great deal of design has been based on reactions involving microorganisms, oxidation, browning, bleaching, vitamin loss, and protein change, so the industry has quickly adopted reaction technology and the use of models to predict and control the shelf life of new foods. The factors that can influence the shelf life of foods are raw materials and product formulation, ambient conditions such as water activity and availability of oxygen, processing and hygiene, packaging materials and system, storage, distribution and display.

The product formulation has a profound effect on the shelf life. Changing the moisture content affects markedly the rates of growth of organisms, browning, bleaching; addition of acids, sugars and salts can affect not only the microbial growth but also physical changes such as crystallisation. By using predictive models, the microbial stability of a product formulation can be rapidly assessed (17). The process can also be designed, using models of inactivation, to ensure that the target microorganisms are effectively destroyed. The packaging design is the link between the design of the product/process and the distribution system. A most important function of food packaging is the protection of the product from environmental conditions, such as light, water vapour, gases and odour, and from internal changes such as loss of moisture, and volatiles. It has to be related to the product qualities and also to the conditions of storage and transport.

The most important of these storage conditions are time and temperature. During distribution, the food will be subjected to different time/temperature regimes. If these are known, then predictive models can be used to follow their effect on the product attributes throughout and at the end of the distribution chain. This can often involve shelf-life trials if there are not sufficient data available. Some steps in design of product shelf life are (18):

- Assess product formulation
- Assess processing conditions
- Specify the product attributes
- Check the history of similar products
- Predict rates of deterioration for critical and important product attributes
- Reformulate product, change process conditions
- Design packaging for product prototype
- Predict storage and transport conditions
- Test shelf life under predicted conditions.

Well-planned shelf-life tests are needed in all product development projects to ensure that the final packaged product is acceptable. There needs to be knowledge of the critical and important product attributes, methods of measuring them, and the conditions to be met in the distribution system, so that the tests can be realistic and applicable. Reaction rate models can reduce the experimental time and work in shelf-life testing.

1.7 Food Processing: Reaction Technology Base

The processing chain in the food system, as shown in Fig. 1.4, is complex and interactive.

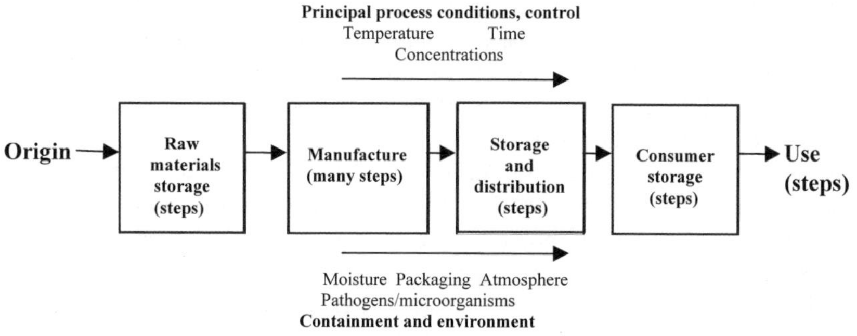

Fig. 1.4. The processing chain

Reactions in one section of the food chain can continue in the next section of the chain; reaction products formed in one section can be the raw materials for reactions in the next section; reactions in a section can also affect other reactions in that section. A problem in developing food processing as a quantitative technology is knowing the reactions and their inter-relationships, and controlling them to give the desired product. Food processing is manipulation of biological materials and therefore has to cope with variation in the raw materials; it is also not an exact process, so there is variability in the processing conditions. The variability may be within set limits but it may still be enough to cause significant changes in the reaction rates in the processing and in the attributes of the final product.

1.7.1 Process variables and their variability

Two groups of variables that affect processed foods are:

- composition of the food, in terms of its components, which can be measured by their concentrations (C)
- processing and distribution environment, principally temperature (T) and time (t), but also other variables, such as those for storage shown in Table 1.II.

Natural variation of the incoming raw materials is an uncontrollable factor that affects processing. These raw materials are almost totally biological in origin and this brings with it dependence on the weather, the time of the year, incidence of contamination and pests, so that there is an inherent variability within materials. An extreme example is the wine industry, which has simply had to accept seasonal variations in the solar insolation and rainfall, which produce, amongst other things, quite major changes in the sugar, acid, and flavour constituents of the resulting grape juice from a given vineyard. This industry has found it quite impossible to accommodate variations overall, although some adjustments can always be made. So the variations are simply passed on to the consumers, who have had to become accustomed to accepting the variation by receiving substantially differing products all with similar labels, often at different prices, and differentiated for the knowledgeable according to the year of the vintage. But this is unusual. In most circumstances, ways have to be found to check out the quality of the raw materials before processing, and then to make appropriate responses in selection, formulation and processing, to ensure a satisfactorily uniform product.

The critical attributes of the raw materials have to be identified, measured and analysed prior to the processing, and then the appropriate treatments or strategies produced to cope with the new, and often still changing, situation. Many of the attribute changes encountered can be subtle and they can also, and frequently do, interact. For example, a reaction can be highly dependent on pH, which can vary in the raw material and can change during processing owing to the chemical reactions occurring in the process. The pH can shift with the quality and the maturity of raw materials, and, where large quantities of these are procured from a number of sources during a growing season, the variations can seem almost random and arbitrary. So all of the relevant factors have to be monitored, and the process conditions adjusted accordingly to maintain consistency in the product. To do this, it is essential to understand as completely as possible the reactions and interactions that arise, their effects on the final food product attributes, and the ways in which the process operators can achieve uniformity in their products.

Processing conditions can also vary. Inevitably in factories under ordinary working conditions there is a variable demand on the services such as the steam supply. Although there is control equipment and the services staff makes every effort, there may still be times when the working steam pressure falls below the stipulated level, or indeed at other times rises above it. These temperature changes give rise to reaction rate changes, and therefore to necessary processing time changes that can be quite dramatic. In Example 1.7, there is a discussion of variation in steam pressure and its effect on reaction rates and therefore on processing.

Example 1.7: Effect of variation of steam pressure on reaction rates in canning

Consider a canning process normally taking place at a fixed temperature of 121 °C, i.e. 2 Bar steam pressure. Time for destruction of the standard microorganism (*Clostridium botulinum*) can be calculated for this temperature using canning calculations. If the absolute steam process pressure were to fall by as little as 9% from 2 Bar, the time needed to

Contd..

Example 1.7 (contd)

accomplish the same extent of processing for destruction of the standard organism would double. The sensitivity of the microorganism to change in temperature works out to be 26% per degree C. Technologists, and perhaps process operators, should easily be able to calculate what changes in process time would be necessitated by a change occurring in their process temperature. And this does not apply just to falls in processing rate; over-processing, though it may be quite safe, will almost inevitably have detrimental effects on quality.

The following questions arise: How can allowances be made for the changing conditions? And if the conditions persist what changes in schedules must be made to compensate? It is the final product that matters, and this is related to the integration of the changes in the food raw material through the processing cycle. How must the variables that are under the control of the operators be manipulated so as to give, reliably, the desired product?

Think break

From Table 1.III, minced beef has a practical storage life of 6 months at -12 °C, 10 months at -18 °C, 15 months at -24 °C.

- If the temperatures throughout your cold store can only be reliably controlled to ±3 °C, and you need to store the meat for 12 months, discuss how you would go about selecting the set temperature for the store.
- If the temperature could be controlled to ±1 °C, how would this affect your choice of store temperature?

1.7.2 Process control

Control of the process can be manual but is increasingly automatic. It is only adequate if:

- sufficient information about the process, including any deviations from the desired levels of the process variables, is available at all relevant times;
- this information can be properly interpreted by either operator or automatic controller;

- the resulting response is sufficient and fast enough to reduce any deviation to a negligible level before the product goes out of specification.

This is an ideal situation to which any real situation only approximates. But modern control equipment with precise and extensive instrumentation can reduce out-of-specification product to a very small fraction of production, so long as the control equipment can follow the processing dynamics adequately and find and ensure a quick corrective response to deviations from the set points. This relies on an adequate understanding, particularly of the process dynamics and also of any interactive possibilities that might upset the situation and in particular lead to over-rapid responses putting either intolerable strains on equipment or instabilities into the processing.

1.7.3 Ensuring product quality and safety in food processing

Control of product quality to within the limits needed for acceptability and marketing of the product is essential for continued access to supermarket shelves and for repeat buying by consumers. These limits are tightening all the time as greater sophistication and greater market power are sought and obtained by both the retail trade and the consumers. Because potential fluctuations in quality have both systematic and also apparently random elements, statistical techniques are the obvious method to use in process control. This has been done over the years with great success by the industry and has yielded real and obvious improvements for the customers.

In the demands made on the industry to reduce any conceivable risk in food products, the conceptions of risk range from the concrete to the vaguest, and at times well beyond anything quantifiable with present knowledge. The 'precautionary principle' basically implies that any incompletely quantifiable risk that exists, however unlikely its occurrence, is unacceptable. Since, from its definition, absence of risk cannot be proved, the only way in which to meet such objections and continue in business is to continually attack ignorance, increase knowledge of reactions in processing, quantify them and reduce their variability.

One powerful way of doing this is by pursuing the components in the food all the way as they move from the raw materials to the consumer product, and then seeking to give the final product a clean bill of health and optimum quality. The clearance must largely come from the lack of harmful outcomes in the historical record. This can be accumulated from information on all components, stemming from analysis and the understanding of what happens in processing. One way of dealing with the danger is to focus on 'food safety objectives', which would define the ultimate level of, for example, pathogens in a given product, and then study the effects of time, temperature and other conditions on the destruction of the pathogens in the process. Although it has considerable practical difficulties in implementation, this use of reaction rates could define the necessary processing that would always ensure a safe product to a stipulated extent or level of risk, as discussed in Example 1.8.

Example 1.8: Ensuring safety in foods

Pathogenic, and toxin-producing microorganisms are a continuing problem in food safety for the food industry, and food safety objectives need to be set. A safety objective would be specific for a particular organism, and would relate to the infective dose, such as, for example less than 100 cfu/g of *Listeria monocytogenes* present in a meat product at consumption.

Meaningful assignment of safety objectives needs to take account of:

- variability of infective dosages and of individuals taking them in
- expense and practicability of meaningful sampling
- probable maximum intake of the food at one meal
- validity of infective dosage as having general applicability.

With bacteriological standards, the problems are substantial, including great biological variation.

To set such a target would need knowledge of the actual *Listeria* input from the raw materials and the environment, followed by reaction technology control of the processing to ensure at most the demanded final levels. Since bacteriological sampling is destructive, time-consuming and expensive, the final outcome would virtually have to be a statistical final risk level at an extent of processing modulated by a combination of Hazard Analysis and Critical Control Point (HACCP) techniques, and final product sampling. The actual risk level achieved would result from the balancing of: sampling and testing adequacy and costs, overprocessing quality losses, raw materials variability, and the statistical risks of infection and reinfection. But it must ensure safety in the final product.

Adapted from Newsome (19)

Think break

In the canning of condensed milk, discuss how temperature could be varied:

* to reduce the time needed to sterilise the milk;
* to decrease the 'cooked milk' flavour in the final product after sterilisation.

And then:

* Discuss how you would organise experiments to determine the time and temperature of canning so that the 'cooked milk' flavour could be at a minimum with the safety of the milk still ensured.

1.8 Food Process Design

The task of the designer of food processing lines and of the component machinery that makes them up is still a mixture of art, science and technology. Even the best designer does not always either get everything right or produce the most efficient and economical plant. The greatest barrier to reducing the gap between what is achieved and what is achievable is lack of knowledge. Food materials and processes are at best complex and only partly predictable, at worst seemingly arbitrary and even wildly capricious. In part, this comes from their biological and fundamentally unstable character, but occurs also in part because technology has not yet reached out and grappled with enough of the problems to solve even most of the serious ones. Computers are taking an increasing part in modern food plant and process design (20), but their effectiveness is never any better than the understanding of the processing built into their programming. In designing a process and equipment, the effects of changing levels of the process variables on the rates of reactions in the food processing and therefore on the attributes of the end product need to be understood and used as a basis for design, as shown in Example 1.9.

Example 1.9: Designing a new process for freezing meat

In order to supply distant markets in Britain, large quantities, hundreds of thousands of tonnes annually, of lamb carcasses have been frozen and exported from New Zealand. Traditionally, carcasses were first cooled in ambient air for about a day and then frozen by placing in static freezer stores until freezing was complete. In the 1950s, air blast freezing was developed and the whole process was speeded up, with the carcasses being loaded during the day into air blast freezers, frozen overnight, and then loaded into the cold stores. This became general in the industry as it saved a good deal of space and time, and it was convenient.

Then reports started to emerge complaining of a greater incidence of toughness, which had not previously been a problem with these young 3- to 6-month-old sheep. A substantial research programme was undertaken and this brought out more of the detail of the muscle biochemistry and of the *rigor mortis* in the tissue. Essentially, it showed that, until certain post-mortem changes in the muscle had occurred, substantial temperature reductions imposed on the muscular tissue could induce shortening of the muscle fibres and this toughened the meat. These changes were chemical reactions, which were temperature-dependent. Control of temperature changes, in particular sufficient time at higher temperatures for the *rigor mortis* to proceed to or near to completion, would avoid the toughening phenomenon. Therefore, prescriptions for time-temperature conditions, which included a stipulated minimum time at higher temperature before

Contd..

Example 1.9 (contd)

exposure to freezing air blasts, became available to the industry from the research and were adopted. This was later extended to include tenderness increases gained by ageing the meat, and combined into a model.

Adapted from Graafhuis *et al.* (21)

Lack of understanding of reaction technology in process design can lead to design mistakes and cause a great deal of money to be lost in redesign, and in drastic cases a loss of a new product for the market. In designing a new process, it is important to:

- identify the critical and important attributes of the final product
- identify the reactions that lead to these product attributes
- identify the process variables, and their possible variation
- determine the effects of changing the process variables on the rates of the reactions
- choose the required levels for the critical product attributes and the optimum levels for the important product attributes
- select limits of the process variables to give the required attribute levels
- design the equipment to give the required control of the process variables.

Obviously, there are engineering and economic needs, which also have to be considered, but these must not disregard the product needs.

1.9 Modelling Food Processing Using Reaction Technology

The food product developer, the plant operator, and the food engineer, in order to work effectively, all need an accurate prediction of what will be produced from the raw materials if they apply defined processing unit operations to them, such as mixing, blending, heating, cooling, ageing, and so on. Over the years, observation has built up a great store of tacit knowledge in the minds of technologists. The tacit knowledge is constantly actively available, consciously and unconsciously, and is always growing. Today, the huge and increasing body of codified knowledge of the food and scientific literature, and of in-house documentation, is adding to this and bringing increasing possibilities of enhanced technology.

One powerful supplement is to have a quantitative description of the changes consequent on processing set out in terms of all of the significant process variables, and this is commonly termed modelling. Figure 1.3 displays the outline for food processing models based on reaction technology. The model of a food process or part of a process is a mathematical description relating the levels of the process variables and the raw materials attributes to the changes in the product attributes. From these relationships, the levels of the product attributes can be predicted for any raw material attribute level and process level, so that an optimum and safe process can be designed. A model is a codified systematic scheme, developed from one situation and then available to be fitted to all sorts of new situations, which obey the same mathematical equations. It is a quantitative description.

The problems are firstly to assemble a model for a specific food process, and secondly to set it up in a form that is conveniently accessible to those who might make industrial use of it. These are almost never researchers, but include operators, managers, designers, and quality controllers. For them, the model has both to exist and to be conveniently available, because, if it is not, then there will be such a barrier to its use that it will be ignored and the benefits not realised. So this means that complicated mathematics, for example, which all too often has to be invoked to deal with the complexities of real food systems, need to be streamlined and included in easily accessible computer software. For example, the canning industry is an area where quantitative information on bacterial destruction has been built up for many years and is presented in simple packages to all operators when they gain their certificates. Aspects of the canning model are discussed in Example 1.10.

Example 1.10: Predicting adequacy of processing in can sterilisation

Attempts to control the safety of heat-processed foods, particularly in metal containers, first started about 80 years ago. There were substantial difficulties in developing quantitative descriptions of, firstly, the effect of heat on microorganisms in various forms and environments and, secondly, the actual heating process in three-dimensional and often complicated container shapes. Simplifications and approximations were based on the use of only one food-poisoning organism, *Clostridium botulinum*, which was heat-resistant, and the slowest-heating part of the can. The contribution to sterilisation of the cooling phase of the heating/holding/cooling process was neglected at first.

Undeniably, if the process determinant is based on the condition at the least processed region of the can, and the process extent varies because of non-uniform heating, there must be overprocessing. Research, over many years, has produced quantitative evidence of the extent of this overprocessing, which can be thought of as a factor of safety. A question then is, does canning need this added factor of safety for its undoubted practical success? Or was it not so much a factor of safety as a factor of ignorance, and it really was retained because of fundamental uncertainties? This would be less important if there were just the question of safety, where

Contd..

Example 1.10 (contd)

there can hardly be too much, but overprocessing also and inevitably detracts from quality as it continues deteriorative reactions beyond the point of necessary processing.

The theory itself, though it has been very useful and demonstrably successful, has several major questions still to be sorted. One, for example, is insistence on 10^{12} reduction in the standard organism at the least processed region in a can – a region that can be small in volume. The vast bulk of the can contents can receive processing greater by orders of magnitude. Since the whole theory is statistically based, there seems justification to extend statistical and averaging treatment over each processed food unit, in this case a can.

Adapted from Hicks (22); Palaniappan & Sizer (23); Peleg & Cole (24)

Think break

For two dried products – instant whole milk powder, and instant meals used by walkers, sailors and other outdoor sportspeople:

* Identify the critical and important attributes of each product
* Identify the attributes in the raw materials related to the critical and important attributes of the products
* Outline the process for each product, showing the unit operations and their sequence
* Identify the nature of changes that occur in the food materials in each unit operation, such as physical change, chemical change, biological change, sensory change
* Relate these changes to the final product attributes you have identified
* Draw basic models, as in Fig. 1.3, to show the reaction technology in the various unit operations in the two processes
* Consider how you might further study the progress of these changes as they proceed with time, what experimental measurements you might need to make, and what sort of results you might expect to come from these experiments.

1.10 The Challenge

To summarise, important issues for the food industry that will be materially advanced by greater understanding of the technological details of food processing, are:

Converting a craft into an industrial technology. This conversion is essential for the food industry to come fully into the modern era and to unlock the potential benefits that the modern technology and the information age can bring with them. It has started, but can usefully be taken much further.

Making use of possibilities, implicit in food raw materials and their interactions, to most effectively and speedily create new products. Only with knowledge and understanding of the scope that exists to change and convert food materials to satisfy needs can the industry fully develop the new products that should be available.

Knowing and understanding and using reaction technology information. There is much available in the literature, but to be used it must be adapted, and extended to calculate processing dynamics that will apply to industrial situations and systems.

Handling and adapting to changes in food materials. Complex change is implicit in biomaterials and biovariability. Therefore, it arises in so many ways in the thriving and dynamic activity that is food processing. Only the fullest achievable knowledge can set up food processors to deal adequately with the continual change that confronts them in every minute, hour, and year.

Building models of food processing. These can be of any available form – descriptive, visual, mechanical, analogue, mathematical – so long as they can be used to make valid predictions about real food materials in industrial situations. Many process variables may be imperfectly understood, even critical ones, but worthwhile progress can still be made, and results can then be improved, as this is needed.

Defining and improving processing equipment. Equipment must operate reliably and be capable of the precision demanded by the quality product. All too often, equipment used in production can be old, out-of-date, sloppy, inadequate, and not capitalising on modern advances that are available. Obvious modern resources include microprocessors and computers, but there also is a whole host of machinery and control developments to be explored for food applications.

Building judgement. Knowledge of processing that is already in the industry needs to be consolidated and extended from the present tacit knowledge that is concentrated in people and their experience towards more general, codified knowledge. Far from diminishing, this will then be available to enlarge the capacity of everyone in the industry and to improve their judgement, both individually and collectively.

Education in processing. Understanding of the fundamentals, and then applying them to industrial problems and situations are key both to improving present operation and to opening the full power and potential of the future.

These challenges illustrate some of the food processing issues that currently challenge food technologists. They are all important, even if only just a representative selection. The application of reaction technology can help to give solutions or partial solutions to the problems they raise, and demonstrate the promise of further experimental and theoretical study. But, more importantly, it is hoped that the outline of the detail that this book provides will convince readers that the methods are both informative and effective in day-to-day operating situations.

Reaction technology should also provide some important and accessible guidelines for making judgements. Although modern knowledge, modern techniques, and in particular computers, have opened up great ease and apparent reliability in process design and control, unfortunately, computers are totally devoid of judgement on particular cases unless judgement itself can be quantified. This can never fully be the case. It is all too straightforward to buy a program, feed in numbers, punch the start button, and then believe and act on the answer that emerges. Technologists are, and need to be, trained to check independently and to think. Empowering thought is where the reaction technology approach can offer accessible tools that are also logical and powerful.

Think break

Information technology is the current new dimension held to offer great and unparalleled advances in food manufacturing operations.

* Relate information technology to the operations of food processing
* How do you see information technology being used to resolve process control and product specification difficulties that currently trouble the food industry?
* How could reaction technology and information technology be combined in process control?
* Outline what further knowledge of the nature of rate processes is needed so that they could be applied in process design and control with information technology.
* Identify food processes where reaction technology and information technology are used to design and control the processes.

1.11 References

1. Earle R.L., Earle M.D. *Food processing: the place for reaction technology*. Leatherhead Food RA Food Industry Journal, 1999, 2 (3), 220-31.

2. Schaffner D.J., Schroder W.R., Earle M.D. *Food Marketing, an International Perspective*. Boston. WCB McGraw-Hill, 1998.

3. Mermelstein N.H. Emerging technologies and "fresh labelling". *Food Technology*, 2001, 55 (2), 22, 64-7.

4. Lewis M.J., Heppell N.J. *Continuous Thermal Processing of Foods: Pasteurization and UHT Sterilization*. Gaithersburg, MD, Aspen, 2000.

5. Earle M., Earle R., Anderson, A. *Food Product Development*. Cambridge. Woodhead, Boca Raton, CRC Press, 2001.

6. Huffman L.M. Processing whey protein for use as a food ingredient. *Food Technology*, 1996, 50 (2), 49-52.

7. Earle M., Earle R. *Building the Future on New Products*. Leatherhead. Leatherhead Food RA, 2000.

8. Hou H., Singh R.K., Muriana P.M. Pasteurization of intact shell eggs. *Food Microbiology*, 1996, 13 (2), 93-101.

9. Polster L.S. *Apparatus and methods for pasteurizing in-shell eggs*. US Patent 6,113,961, 2000.

10. Singh R.P. Scientific principles of shelf life prediction, in *Shelf Life Evaluation of Foods*. Edited by Man C.M.D., Jones A.A. London. Blackie Academic and Professional, 1994, 3-27.

11. Ellis M.J. The methodology of shelf life prediction, in *Shelf Life Evaluation of Foods*. Edited by Man C.M.D., Jones A.A. London. Blackie Academic and Professional, 1994, 27-39.

12. Van Arsdel W.B., Copely M.J., Olson R.L. (Eds) *Quality and Stability of Frozen Foods, Time-Temperature Tolerance and its Significance*. New York. Wiley-Interscience, 1969.

13. Erickson M.E., Hung Yen-Cong (Eds) *Quality in Frozen Food*. New York. Chapman & Hall, 1997.

14. International Institute of Refrigeration. *Recommendations for the Processing and Handling of Frozen Foods, 3rd Edition*. Paris. International Institute of Refrigeration, 1986.

15. International Institute of Refrigeration. *Storage Lives of Chilled and Frozen Fish and Fish Products*. International Institute of Refrigeration Proc. Comms. C2 and D3. Paris. International Institute of Refrigeration, 1985.

16. Earle R.L. Quality loss rates in chilled foods, in *Technologies for Shelf-Life Extension of Refrigerated Foods Workshop*. Joint Australia and New Zealand Institutes of Food Science and Technology Conference, Auckland, 1995.

17. Walker S.J. The principles and practice of shelf life prediction for microorganisms, in *Shelf Life Evaluation of Foods*. Edited by Man C.M.D., Jones A.A. London. Blackie Academic and Professional, 1994, 40-50.

18. Jones H.P. Ambient packaged cakes, in *Shelf Life Evaluation of Foods*. Edited by Man C.M.D., Jones A.A. London. Blackie Academic and Professional, 1994, 179-200.

19. Newsome R.H. Science, communication and government relations. Reporting a verbal quote from Hodges J. of the American Meat Institute Foundation. *Food Technology*, 2001, 55 (2), 20.

20. Datta A.K. Computer-aided engineering in food process and product design. *Food Technology*, 1998, 52 (10), 44-52.

21. Graafhuis A.E., Lovatt S.J., Devine C.E. *A predictive model for lamb tenderness*. Proceedings of the 27th Meat Industry Research Conference. Hamilton. Meat Industry Research Institute of New Zealand, 1992.

22. Hicks E.W. Some implications of recent theoretical work on canning processes. *Food Technology*, 1951, 5, 175-8.

23. Palaniappan S., Sizer C.E. Aseptic process validated for foods containing particulates. *Food Technology*, 1997, 51 (8), 60-8.

24. Peleg M., Cole M.B. Reinterpretation of microbial survivor curves. *Critical Reviews in Food Science and Nutrition*, 1998, 38 (5), 353-80.

2. PRODUCT CHANGES DURING PROCESSING

2.1 Introduction

Processing converts the food raw material into the food product. The process must be designed to ensure that, in the final product, the composition and the levels of the other product attributes are as demanded in the specification. Processing moves in steps towards this end product. The final qualities are therefore created through the processing steps, by reactions causing changes in the food materials. These changes have to be instigated; they have to be continued at rates that are as fast as can be operated industrially, yet they must be kept under control. Then they must be stopped, when the necessary qualities in the final product have been reached but before they are exceeded. To do all these things, the best attainable understanding is needed of the:

- raw materials – composition and other attributes inherent in the food raw materials;
- reactions – causing changes in composition and other attributes of food raw materials during processing;
- processing conditions – external factors which instigate and modify these changes;
- final product – composition and other attributes specified in the final product.

The composition and the attributes of the raw materials and the possible changes are basically a matter of chemistry. But foods are complex chemical systems, and, in reaction technology, other physical, sensory and biological changes are followed as well as chemical changes. The rates of changes, their inception, modification and control, need to be studied and understood so that they can be implemented in manufacturing practice. This chapter takes an initial look into the rates of individual food processing reactions and the effects of some controllable variables, particularly processing temperature, and constituent concentration, on them.

The rates at which changes occur can be understood, predicted and controlled, and this knowledge can be applied to design and control the process to give the specified final product quality. A general scheme for the use of reaction technology in processing is shown in Fig. 2.1.

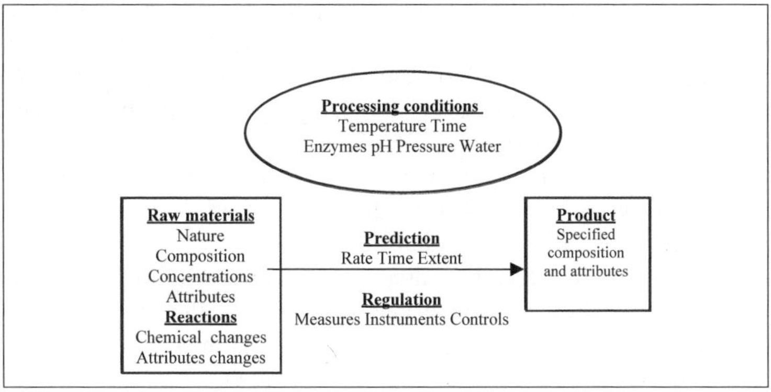

Fig. 2.1. Reaction technology in processing

2.2 Reactions in Food Materials During Processing

An immediate impact can come from focusing on a practical example of processing, preferably a simple one. One common and relatively uncomplicated process is the manufacture of jam by boiling fruit and sugar. It is a simple and straightforward process, encountered domestically as well as industrially, and yet one that illustrates very many of the important principles that arise in processing and which have to be taken into account to arrive at a good-quality product. In jam processing, many reactions have to take place before the ingredients are changed into the food product.

The ingredients are generally only fruit pulp and sugar, to which pectin and acid may have to be added if not present in sufficient concentration in the fruit, and perhaps additional water for handling. The reactions start on heating the pulp and sugar mixture; effectively, they do not occur without this heat. So the first point emerging is that reactions are initiated and speeded by higher temperatures. Some reactions in jam making are shown in Table 2.I.

TABLE 2.I
Some reactions in jam making

Reactions		Products
Sucrose hydrolysis catalysed by acid	→	glucose and fructose
Sucrose hydrolysis catalysed by enzymes	→	glucose and fructose
Caramelisation/burning of sugars	→	darkening and caramel/burnt flavours
Browning, sugar/protein (Maillard) reactions	→	darkening, bitter flavours
Colour bleaching	→	colourless compounds
Pectin polymerisation	→	gelation
Pectin breakdown catalysed by enzymes	→	simpler carbohydrates
Enzyme activation/inactivation	→	increase or stopping of reactions

Sugar hydrolysis, where the acid and enzymes in the fruit cause the sugar (sucrose) to undergo hydrolysis (inversion) into glucose and fructose (invert sugar). High concentrations of dissolved solids in the jam are needed (65-70% total solids content) to create high osmotic pressures and reduce water activity so that potential spoilage organisms can no longer grow. Therefore, total sugar concentrations need to be high, but, if high enough in sucrose, its crystallisation can occur. So some inversion of the sucrose, in fact about 50%, is used to prevent its crystals from forming in the final product. The invert sugar contributes to the high concentrations but does not crystallise.

Pectin changes, in which carbohydrate polymers present mostly in the peel of the fruit polymerise further to give gels that solidify the jam. But enzymes, pectinesterases, may also break down the pectins and interfere with the gelling needed to give substance to the product. So there are enzymes that are helpful in aiding the sugar hydrolysis, and enzymes that are unhelpful in destroying pectin functionality.

Sugar caramelisation, where the sugars caramelise or in more extreme circumstances start to burn, forming dark colours and unwanted flavours. The reducing sugars can react with proteins that are present, through the browning (Maillard) reaction.

Complex flavour and aroma constituents in the fruit can evaporate and be lost, but they can also react, generally to form less desirable products. Colour constituents can bleach and change.

So there are reactions, occurring simultaneously, in parallel, and in sequence. There may be interaction – for example, the pH may change, and the enzymes may be inactivated by internal reaction to a greater or lesser extent, and caramelisation can be affected by the degree of sucrose hydrolysis.

Think break

Choose a food manufacturing system, e.g. bread baking, or blanching and freezing vegetables.

* Identify the critical and important constituents and attributes of the raw materials
* List the reactions that are being induced in the processing, the constituents involved both as reactants and as products, and the attributes being changed
* Identify the processing conditions causing the changes
* Identify the critical and important constituents and attributes in the final product.

2.3 Time and Temperature in Food Processing

Time is a critical variable in food processing. At this point it may be helpful to look at the range of time spans that are encountered in food processing, to put time scales into perspective. Food processing occurs on a time scale from a few seconds in a rapid heating process to years in shelf lives on storage. The rates of the important reactions in the process vary, respectively, from very fast to very slow. In using reaction technology, times are easier to compare with all of them in the same units, usually minutes, even though the numbers can commonly vary from 10^{-1} to 10^{6} min. Because we shall be looking at processing rates, the average rates can be compared by looking at the reciprocal of the time needed for the process to be completed. Thus for the short time scale for manufacturing/ processing, of between 0.1 and 100 min duration, rates vary from 10^{1} to 10^{-2} min^{-1}; for preservation lasting from a few days to a few months, average rates vary from 10^{-4} to 10^{-6} min^{-1}. And so the corresponding practical 'half-life' range, the time needed for 'concentrations' to halve, is a wide one, from around 10^{-1} to around 10^{6} min. Times and the equivalent rates of reaction are shown in Table 2.II.

TABLE 2.II
Processing times and rates of reactions

Time (min)	Rate of reaction (min^{-1})
0.01 (0.6 s)	10^{2}
0.10 (6 s)	10^{1}
1.0	10^{0}
10	10^{-1}
100 (1.7 h)	10^{-2}
1,000 (17 h)	10^{-3}
10,000 (7 days)	10^{-4}
100,000 (70 days)	10^{-5}

Some practical examples are:

- Seconds to 1 min: very short-time flow heat sterilisations, rates 10^{1} to 10^{0} min^{-1}
- 10 min to 1 hour: heating, cooking, canning, baking, rates 10^{-1} to 1.6×10^{-2} min^{-1}
- 2 hours to a day: curing meat, rates 8.3×10^{-3} to 6.9×10^{-4} min^{-1}
- 10 days to 2 years: ambient, chilled and frozen storage, maturation, rates 7×10^{-5} to 1×10^{-6} min^{-1}

Reactions such as denaturation of proteins, gelation and hydrolysis can have rates varying from 10^{-1} to 10^{-3} min^{-1} depending on process conditions.

Temperature is another critical variable. Temperatures lie over limited ranges, running from approximately 250 °C (or about 523 K) at the high end, briefly on the surface of foods in an oven, down to -25 °C (or about 248 K) in freezer stores. There are a few and substantially more expensive cold storage temperatures down to as low as -70 °C (200 K) for very special food situations but these are rare. So the working range for most food reactions is essentially from +250 °C to -25 °C, or from 523 K to 248 K, i.e. a range of 275 degrees. For many purposes, Celsius units (°C) are convenient and widely understood, but sometimes (although the unit intervals stay the same) it is necessary to move to the Kelvin scale (K) based from absolute zero (-273 °C), and therefore with K = (°C + 273). Processing and storage temperature ranges are shown in Table 2.III.

TABLE 2.III
Processing and storage temperatures: ranges for typical food processing reactions

Process/storage	Temperatures (°C)	Temperatures (K)
Frozen storage	(-70)-1	203-272
Chill storage	0-5	273-278
Ambient processing	10-30	283-303
Warm processing	40-80	313-353
Hot processing	80-150	353-423

2.4 Concentration Sensitivity

The rate at which a food material is changed in a reaction, in terms of its mass transforming with time, has been found to relate to the mass itself that is present and so to its concentration. Concentration measures the closeness of the molecules together and thus their potential to react with each other. One reaction system that can be used to study the effect of concentration on rate of reaction is the hydrolysis of sucrose to invert sugar. Sucrose hydrolysis is a simple chemical reaction, and shows how the effect of concentration on rate of reaction can be quantified.

2.4.1 Rate of change proportional to concentration

The rate of change in sucrose has been found experimentally to depend directly on the sucrose level present, that is on the sucrose concentration. Therefore, as the reaction proceeds and the sugar concentration drops, so does the rate of the reaction.

The rate of change of sucrose to invert sugar is:

$$\text{Rate of reaction } (r) = dC/dt$$

where C is the concentration, t is increasing time; and dC is the decrease in concentration during the time interval of dt.

Because the rate of the reaction is proportional to the concentration of sucrose, an equation can be written:

$$r = dC/dt = - k\, C$$

where r is the rate of the reaction, k is a proportionality called a reaction rate constant, and C is the concentration of sucrose, the quantity of sucrose present in a unit of volume. The negative sign arises because the concentration of the reactant C decreases as time t increases. In this case the concentration of the sucrose could for example be measured in grams per litre, g/l.

This form of dependence is called a ***first order reaction***, and is one that is encountered commonly throughout food processing. The dependence of rate of reaction on concentration can more generally be expressed by proportionality to some power of the concentration,

$$r = dC/dt = - kC^n$$

but in the case of sucrose inversion it has been found to be linear, that is concentration (C) is raised to the first power. Other orders, zero and second order, are described in Theory 2.1. Many food processing reactions only approximately follow such simple equations over the full range of the reaction, but this particular reaction follows it throughout. It has been very extensively studied and it also just happens to be needed in making jam. It is a classical chemical reaction, and it has been experimentally found to fit the equation very well.

2.4.2 Time needed to reach a particular concentration

To predict the time needed to move from the concentration of sucrose in the raw materials to that of the final jam, the equation is integrated. The reaction rate constant can be determined experimentally, or in some cases the data can be found in published reports. In Example 2.1, the results for sucrose hydrolysis are taken from an old data handbook (1), which makes the additional point that information useful in the food industry is often available from a wide variety of sources. It is worth noting that the range over which the concentrations have been measured (1,000-fold ratio) is wider than might normally be determined in an industrial study.

Having arrived at the rate equation, then the integration, which is just summation over time, allows either the time to be found knowing end concentrations, or the end concentration to be found for a given time of reaction. The mathematics of the integration is shown in Theory 2.1, where results are also given for reactions of other orders.

Theory 2.1: Integration of rate equations

The equations can be integrated formally, keeping to one component.

General equation $dC/dt = -kC^n$ where n is the order

First order n = 1 $dC/dt = -kC$
and so $dC/C = -k\,dt$
and integrating $\ln C/C_0 = -k(t-t_0) = -k\,t$ if $C = C_0$ when $t = t_0 = 0$
i.e. $\ln C/C_0 = -kt$

General order n (n≠1) $dC/dt = -kC^n$
and so $dC/C^n = -k\,dt$
and integrating $\{C^{1-n} - C_0^{1-n}\} = (n-1)\,k\,t$

For zero order when n = 0, then

$$C - C_0 = -k\,t$$

For second order when n = 2

$$C^{-1} - C_0^{-1} = 1/C - 1/C_0 = k\,t$$

Many practical purposes can be covered by zero and first order, and sometimes second order is added. The addition of other and fractional orders, especially over two, is rarely needed. Further information on the analysis of reaction rates can be found in Levenspiel (2).

This leads to the integrated equation for *first order*:

$$\boldsymbol{t = (-1/k_T)\ln(C/C_O)}$$

In this equation, t is the time needed for the concentration of the sucrose in the jam, in for example g per litre, to move from its initial value C_O to a final value of C, and k_T is the reaction rate constant at temperature T. The reaction rate constant, k_T has units of inverse time (commonly min^{-1}) subscripted T to indicate the temperature at which it applies, and ln indicates taking the logarithm to base e (found in practice by keying in the number and then pushing the ln button on the hand calculator). The negative sign comes in from the integration (you will find a problem if it is not included as C/C_0 is less than 1 and so $\ln(C/C_0)$ is negative). The equation may look a little formidable, but it can be worked through using a hand calculator or on a computer spreadsheet. This is shown in Example 2.1, illustrating hydrolysis (inversion) of sucrose.

Example 2.1: Sucrose: change of concentration with time (hydrolysis)

The times needed for inversion of sucrose in N/100 HCl at 80 °C, are given as 50% inverted after 9.1 min, 90% after 30.3 min and 99.9% after 90 min (1).

These values are plotted in Fig. 2.2(a). Figure 2.2(a) shows the direct plot of C/C_0 against time and on to this a computer-generated trend line has also been added.

Then the logarithms of C/C_0 are calculated.

Time (min)	9.1	30.3	90
C/C_0	0.5	0.1	0.001
$-\ln(C/C_0)$	0.693	2.303	6.91

C/C_0 are plotted on a logarithmic scale against time in Fig. 2.2(b).

Figure 2.2(b) gives a logarithmic plot showing an excellent straight line, and so giving positive confirmation to a first order reaction.

The reaction rate constant at 80 °C (k_{80}) can then be found from the equation:

$$t = (-1/k_{80}) ln(C/C_0)$$

as it is the slope of the line on the logarithmic graph.

$$k_{80} = 0.076 \text{ min}^{-1} = 7.6 \text{ x } 10^{-2} \text{ min}^{-1}$$

The equation can be used to determine the time for any ratio of C/C_0 and also the concentration ratio for any time at 80 °C.

This example shows how once the rate of reaction is determined from sufficient experimental results, the equation can be used to determine the time for any specified concentration of sucrose. This time is obviously a key aspect for the jam manufacturer, determining the time needed for the processing and, amongst other things, the throughput of the equipment and the output from a particular process line. It can also provide a time signal for the operator to move from general to close supervision as the desired end point is approached. This may not explicitly be needed by a skilled operator but is a help for the less skilled and can be worked into the control strategy as the process control moves towards automation.

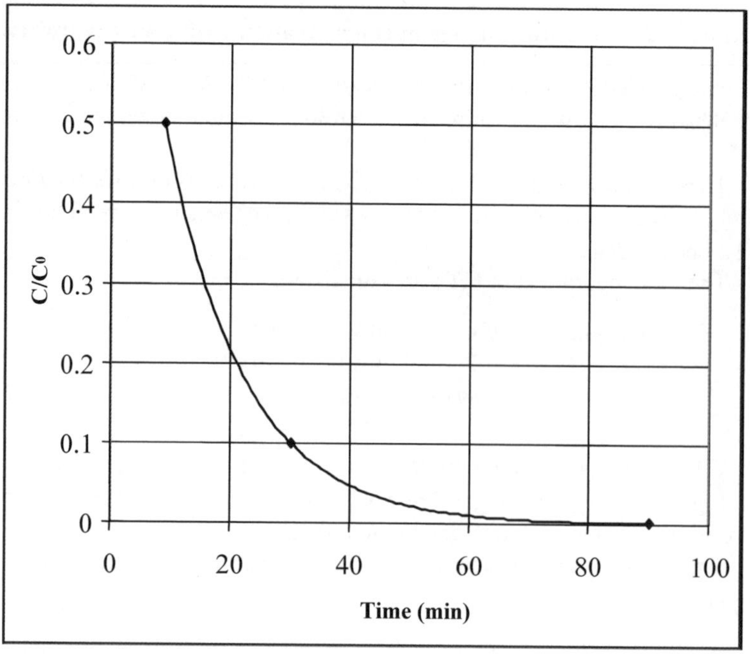

Fig. 2.2(a). Sucrose hydrolysis at 80 °C – natural scale

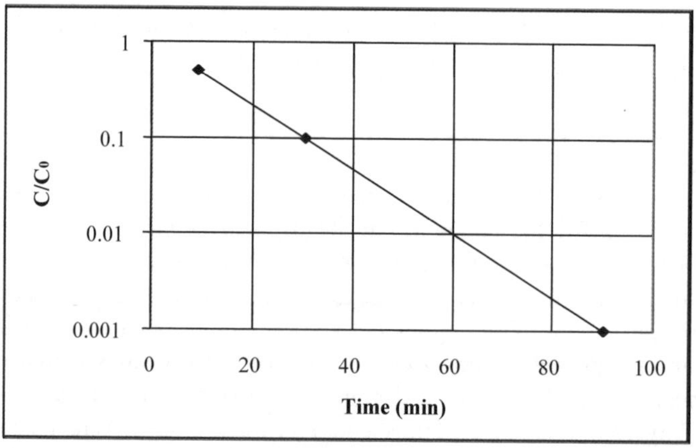

Data from: International Critical Tables (1)

Fig. 2.2(b). Sucrose hydrolysis at 80 °C – logarithmic scale

Think break

Using the information in Example 2.1:

* work out intermediate sucrose concentrations at times: 10, 20, 50, 70 min.
* work out the times for different concentrations: 60, 70, 80, 90%.

For example, after 30.3 min, $C/C_0 = 0.1$ so $\ln (C/C_0) = -2.303$ and therefore

$$k_{80} = -(1/t)\text{x} \ln (C/C_0) = -(1/30.3) \text{ x } (-2.303) = 0.076 \text{ min}^{-1}$$

Using the equation

$$t = (-1/k_T) \ln (C/C_0)$$

After 15 min, the sucrose concentration can be calculated:

$$C/C_0 = \exp (-0.076 \text{ x } 15) = 0.32$$

that is concentration has dropped to 32% of the original concentration. The time taken to reach 50% of the original concentration, $C/C_0 = 0.5$

then $t = -(1/0.076) \text{ x } \ln(0.5) = -13.2 \text{ x } (-0.693) = 9.1$ min.

2.4.3 Rate equations

It may be appropriate to notice here the ways in which measures of concentration in food processing can take many forms, and these include mg/ml, g/l, kg/tonne, g/100 g, mol/l, microorganisms/cm^2 surface or per cm^3 volume, and so on. Similar reaction rate equations have also been found to fit some measures of consumer acceptance. This diversity of measures conforms to kinetic rate equations so long as the measures are quantitative. With only a change in numerical constants to make the differing numbers fit consistently, the pattern stays the same.

The rate may also be proportional to the concentrations of other components, C_B, C_C, and so on, but for the moment it is best to concentrate on one only, just $C_A = C$, in our example sucrose. The others will follow similarly.

When looking at rate equations it is always worth remembering that, from the present point of view, that of the process technologist, rate equations are simply descriptive. It is a matter of finding an equation that fits the experimental data, and then using it only so long as it continues to fit closely enough for the technical purpose in hand. There may also be further systematic and theoretical ways of looking at the systems that may be helpful, but they are not necessary for technological calculations and predictions, so we need not worry about them here in order to reach the required end product satisfactorily.

Think break

- Using a hand calculator, arbitrarily select a value for the reaction rate constant k (perhaps 1 min^{-1}) and an initial food constituent concentration (perhaps 300 g/l), and, assuming a first order reaction, calculate progressive concentration/time values until the concentration reduces to 1% of its initial value.
- Plot the concentrations linearly and logarithmically, against time.

As an example for $t = 0.5$ min and $C_0 = 100$ g/l, with $k = 1$ min^{-1} then

$$t = (-1/k_T)\ \ln (C/C_O)$$

$$0.5 = (-\ 1/1)\ln (C/100) = -1\ \ln(C/100)$$

therefore $C = 100\ \exp(-0.5) = 100\ e^{-0.5} = 61$ g/l

(If unfamiliar with exponentials, run these numbers back and forth a few times on a calculator to convince yourself that they all fit together and make sense.)

The integrated first order equation can be applied under any concentration circumstances. For example, when the concentration halves C becomes $C_0/2$ and so

$$\mathbf{-ln(C/C_0) = -ln(½) = 0.693}$$

$$= k \text{ x (time for concentration to halve)} = k \text{ x } (\textit{half-life}) = k\ t_{0.5}$$

This *half-life* gives some feeling for the magnitude of the reaction rate constant because k for these first order reactions equals {0.693/(the half-life)}. With the logarithmic nature of the first order reaction, in a further time interval of a half-life the concentration will halve again, and so on. So, for example, if our sucrose hydrolysis has a half-life of 9.1 min, in a further 9.1 min it will halve again, that is to ¼ of its original concentration, and for the reaction discussed in Example 2.1 at 80 °C, the reaction rate constant is 0.693 $(1/t_{0.5})$ = 0.693 (1/9.1) = 0.076 min^{-1} as set out in the Example.

The concept of a reaction rate that is linearly related to the present concentration, and which therefore generates constant fractional, first order reaction rates, is in fact a rather fundamental one. What it says for the hydrolysis in the jam is that the initial rate at which the sucrose that was added to the fruit starts to invert, that is to change into glucose and fructose, is proportional to the actual amount of the sucrose that was added. And further, as the sucrose reacts and its concentration decreases, so the reaction rate in terms of, say, grammes of sucrose inverting per minute decreases proportionately. This seems a rather probable and intuitive behaviour. So it is hardly surprising that it has many wide-ranging applications.

Processing times are critically important in food processing. We have already encountered times for concentration changes. Another very common instance arises in both the growth and the death of microorganisms. In growth patterns in biology, the concept of a doubling time is well known. It is just another example of increase being based on present numbers (concentrations), applied in this case to cell division. Microorganism growth is important in food processing for two broad reasons: for the building up of desired cultures to establish flavours and textures, and in the unwanted proliferation of pathogens and spoilage organisms creating toxins, potential infections, off-flavours, less desirable textures, colours and slimes. The death/inactivation of microorganisms follows a similar pattern during processing, at least over limited ranges. Numbers of microorganisms can be reduced dramatically, thus decreasing deterioration in the food. Heating is one way of deliberately inducing death by increasing local energy intensity. Other methods of increasing local energy intensity are irradiation, much elevated pressures, and electric fields. After such treatment, the number of live organisms, or their ability to metabolise and reproduce, decreases and it has been found to do so systematically.

2.5 Temperature Sensitivity

As well as having concentration sensitivity, reaction rates are also sensitive to temperature. A problem in processing is that the temperature may be not constant. In our continuing example, during processing, the jam has to be heated up and cooled down. If steam pressures are not steady, then heating jacket temperatures can vary accordingly, and also as the jam is boiled its solid content rises and so does its boiling temperature. Changes in temperature are implicit in most of processing. Reaction rates generally increase with increasing temperature, and the reaction rate constant is only constant as long as the temperature is also. So the reaction rate constant has also to be predictable in terms of the working temperature for it to be useful in processing. This can be done.

2.5.1 Relationship between reaction rate and temperature

Much experimentation has shown the rate/temperature relationship for a simple irreversible reaction to be an exponential one in which the basic rate equation is expanded and of the form:

$$-r = k_T C^n = (\mathrm{A}\ e^{-E/RT})\ \mathrm{C}^n$$

and, on cancelling the concentration terms:

$$k_T = A\, e^{-E/RT} = A \exp\{-E/RT\}$$

This is a fundamental equation connecting reaction rate constants with absolute temperature. Here r is the rate of the reaction, k_T *is* the reaction rate constant and subscripted T to indicate that it is a function of temperature, A is called the ***frequency factor*** and is a constant for a particular reaction, E is called the ***activation energy*** of the reaction (in Joules/mol.), T is the temperature (absolute, in degrees Kelvin, K = 273 + C) at which the reaction takes place, C is the concentration of the reacting component, and R is a constant (called the 'universal gas constant' and numerically 8.314 Joule/mol. K).This shows that rates and temperatures are connected through an exponential (exp), and further that the temperature occurs in the exponent (the term in the {} brackets) as the reciprocal (1/T) of the temperature measured in degrees K.

This is commonly named the ***Arrhenius equation*** after its originator, who, well over 100 years ago and working as it happens on the sucrose hydrolysis system, advanced the idea. His choice must have been sound as it has remained the best for many purposes ever since. An enormous amount of work has been done on it, and, while never definitively proven, no simple experimental reaction has ever been found that departs from it (3). Its form may look complicated, but once more the hand calculator comes to the rescue when working in practice.

The Arrhenius equation can also be written:

$$\ln k_T = \ln A - (E/R) \bullet (1/T)$$

In the ***Arrhenius plot***, the natural logarithms of the reaction rate constants at different temperatures are graphed against the reciprocal of temperature(1/T). If the data conform to the equation, then the graph is a straight line and the value of E/R is the slope of the line. As R is constant, the activation energy, E, can be found. This is shown in Example 2.3 for the hydrolysis of sugar.

The Arrhenius equation and its implications are at the heart of food processing reaction technology because so often temperature is the primary agent in initiating and controlling the actual processing. It contains two terms, one a multiplier, A, and the second an exponential, $e^{-E/RT}$, which can also and perhaps more conveniently be written $\exp\{-E/RT\}$. The magnitudes of the terms A and E once known, for any particular reaction, allow calculation of the rates at different temperatures, T_1, T_2, and so on. The first, A, the frequency factor, in effect sets the general level of the rate and is often found to be numerically very large (for example of the order of 10^{10}). Everyday manageable processing levels of the rates (from about 10^{-3} min^{-1} to about 10 min^{-1}, implying half-lives from about 10 hours to a few seconds) emerge only after it has been multiplied by the second term. The second term has a negative exponent, and is often in real circumstances found to be extremely small (for example of the order of 10^{-10}). It contains in the exponent the quantity E, the activation energy, and this characterises classes of reactions. It has the effect of setting the sensitivity of a reaction to (absolute) temperature T. The term R is there for thermodynamic reasons only; it is a constant. Working with this equation, the changes in reaction rates consequent upon a given change in temperature can be calculated. Also sensitivities, which are the percentage change in reaction rate due to unit change in temperature, can be calculated. These can be so useful when considering rates in practice.

Reaction rate constants at different temperatures are connected by:

$$k_{T1} / k_{T2} = \exp \{E/R(1/T_2 - 1/T_1)\}$$

found by dividing the Arrhenius equation for the one temperature by that for the other.

The time for 50% hydrolysis at temperature T can be determined by using:

$$t = (1/k_T) \ln ½ = 0.693 (1/k_T)$$

Example 2.2 illustrates the effects of temperature on reaction rate constants in hydrolysis of sucrose.

Example 2.2: Sucrose hydrolysis: reaction rate constants at different temperatures

The reaction rates at different temperatures for the hydrolysis of 50% sucrose solution in 0.1N HCl are given in the International Critical Tables (1).

Temperature (T)						
	(°C)	0	15	30	40	50
	(K)	273	288	303	313	323
1/T	(1/K)	3.67×10^{-3}	3.47×10^{-3}	3.30×10^{-3}	3.19×10^{-3}	3.10×10^{-3}
Reaction rate (k)						
k	(min^{-1})	7.7×10^{-6}	9.2×10^{-5}	8.7×10^{-4}	3.3×10^{-3}	1.2×10^{-2}
ln k		-11.77	-9.29	-7.05	-5.71	-4.42

ln k is plotted against 1/K in Fig. 2.3, the Arrhenius plot for sugar hydrolysis.

Contd..

Example 2.2 (contd)

This is a straight-line relationship, so the slope gives the value of:

$$E/R = (11.77 - 4.42)/(3.67\text{x}10^{-3} - 3.1\text{x } 10^{-3}) = 7.35/0.57\text{x}10^{-3}$$
$$= 12.9\text{x}10^{3} \text{ K}$$

R, the gas constant, is 8.314 joules/mol K

$$\text{so E} = 107\text{x } 10^{3} \text{ J/mol}$$
$$= 107 \text{ kJ/mol}$$

To estimate the reaction rate at 110 °C, that is 383 K,

$$k_{383} / k_{323} = exp\{-E/RT_{383}\} / exp\{-E/RT_{323}\}$$
$$\text{so } k_{383} / 1.2\text{x}10^{-2} = exp\{E/R\ (1/323 - 1/383)\}$$
$$= exp\ (12.99\text{x}10^{3}(3.096\text{x}10^{-3} - 2.611\text{x}10^{-3})$$
$$= exp\ (6.30)$$
$$= 545$$
$$\text{and so } k_{383} = 545 \text{ x } 1.2\text{x } 10^{-2}$$
$$= 6.54 \text{ min}^{-1}$$

To estimate the time for 50% hydrolysis at 110 °C, that is 383 K

$$-t_{0.5} = (1/6.54)(\ \ln \tfrac{1}{2})$$
$$= 0.153 \text{ x } -0.693$$
$$= 0.11 \text{ min}$$

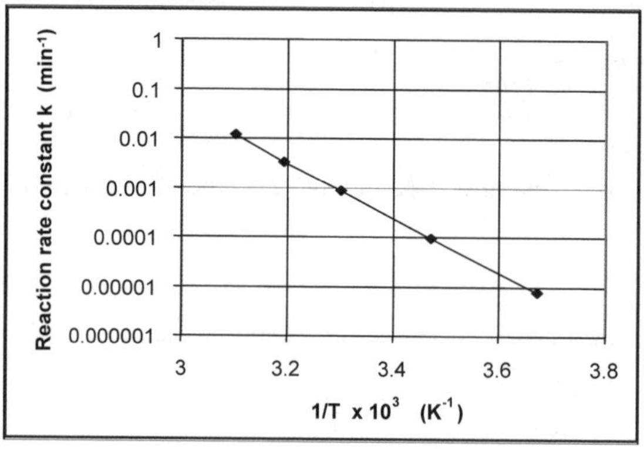

Fig. 2.3. Sucrose hydrolysis – Arrhenius plot for reaction rate against temperature

Think break

The reaction rate constants, k, for the hydrolysis of sucrose at temperatures from 60 to 91 °C using 0.1NHCl are:

Temperature T (°C)	59.90	69.97	80.13	90.29	90.32
Reaction rate constant k (min^{-1})	0.4000	0.1236	0.3687	1.033	1.020

* Draw the Arrhenius graph from these data for the hydrolysis of sucrose between 50 and 90 °C
* Determine the activation energy E
* Predict the reaction rate constant, k, at 100 °C
* Determine the time for 50% hydrolysis at 90 and 100 °C
* How do the reaction rate constants vary with increasing temperatures?
* What does this mean for process control?

2.5.2 *Sensitivity to processing temperature*

The sensitivity of processing to temperature can be worked out very straightforwardly by remembering that it is from $(k_{T+1})/(k_T)$ = (1 + sensitivity) or it can be related to the activation energy, as shown in Theory 2.2.

Theory 2.2: Temperature sensitivity calculation

The Arrhenius equation is

$$k = Ae^{-E/RT}$$

Taking the derivative $dk/dT = (Ae^{-E/RT}) \bullet E/(RT^2) = k\,[E/(RT^2)]$

Therefore $dk/kd(T) \doteq d(\ln k)/dT = E/RT^2$

And since $d(\ln k)/dT \doteq \ln[(k_{T+1}) / (k_T)]$

Therefore $E/(RT^2) \doteq \ln[(k_{T+1}) / (k_T)]$

$$\exp(E/RT^2) \doteq (k_{T+1}) / (k_T) = 1 + \text{sensitivity}$$

or $$\exp(E/RT^2) - 1 \doteq \text{sensitivity}$$

Because:

$$\mathbf{(k_{T+1})/(k_T) = \exp\{E/RT^2\} = 1 + sensitivity}$$

R is a constant and 8.314 J/mol K, for a reaction with an activation energy of, say, 100,000 J/mol, the sensitivity at 100 °C or 373 K, can be found from:

$$\exp\{100{,}000/(8.314.373.373)\} = 1.09 = 1 + (9/100)$$

This (9/100) can be thought of as a temperature sensitivity of the reaction rate amounting to 9% per degree C and written 9%/°C. Experimentally, 9%/°C is found to be close to the temperature sensitivity of the sucrose hydrolysis reaction. In case you are not familiar with exponentials, it is salutary to consider their power in these ratios, as shown in Example 2.3.

Example 2.3: Demonstrating the power of logarithmic temperature sensitivities

Sensitivities seem at first sight to be rather innocuous, but their "logarithmic" force gives them great power. For example, with a 26% per degree sensitivity:

- a 3 °C shift upward doubles the rate of the reaction (1.26^3) = 2;
- a 5 °C shift increases the rate (1.26^5) = 3.2 times, and
- a 20 °C shift multiplies it (1.26^{20}) = 102 times.

While a 20 °C shift may seem large around 100 °C, it is not an altogether remote possibility in the fluctuations of an oven at 200 °C.

Think break

For three activation energies: 100, 200 and 400 kJ/mol

* Work out the sensitivities at temperatures of 100 °C and 110 °C.
* Determine the ratio of reaction rate constants, k_{110}/k_{100}.
* Reflect on the magnitudes of these numbers and of their significance in a food process in which they might occur.

For example, for E = 200,000 J/mol at 100 °C = 373 K,

$$E/RT^2 = 0.173$$

so $\exp\{E/RT^2\} = 1.19 = (1 + \text{sensitivity})$, so the sensitivity is 19%/ °C.

Over a 10 degree temperature range

$$k_{383}/k_{373} = A\exp(-E/R.383)/\ A\exp(-E/R.373) = \exp\{E/R(\ 1/373 - 1/383)\}$$
$$= \exp\{\ (200{,}000/8.314)\ (1/373 - 1/383)\} = \exp\{\ 1.684\} = 5.4$$

so this reaction moves over five times faster at 110 than at 100 °C.

2.6 Reaction Rate/Temperature Relationships: Activation Energies

Food processing predictions can be made using the ranges of concentration and temperature found in food processing combined with the levels of activation energies calculated from experimental observations. The activation energies broadly lie between those of many chemical reactions, with a minimum of about 50 kJ/mol and an average probably nearer to 100 kJ/mol, and those of protein denaturations (and bacterial deaths that may be caused by somewhat the same basic molecular rearrangements) with a maximum of about 500 kJ/mol the activation energies of many enzyme reactions lie in between these. The total range is about tenfold, 50-500 kJ/mol.

Activation energies for particular reactions have to be determined experimentally. This is normally done by measuring rates over a range of temperatures and then constructing the so-called Arrhenius plot of the logarithm of the measured reaction rate constants against the reciprocal of the absolute temperature, (ln*k*) against (1/T) as illustrated in Fig. 2.3. Table 2.IV gives a general picture of the expected magnitudes, together with the calculated sensitivities at two temperatures, using the higher of the activation energy ranges.

TABLE 2.IV
Activation energies for typical food processing reactions

	E	**Sensitivity***	
	kJ/mol	**(40 °C)**	**(120 °C)**
Chemical reactions			
General chemical reactions	40-100	13%/ °C	8%/ °C
Hydrolysis reactions	60-120	16%/ °C	10%/ °C
Lipid oxidations	40-100	13%/ °C	8%/ °C
Browning (non-enzymic) reactions	100-200	28%/ °C	17%/ °C
Vitamin destruction	70-150	18%/ °C	12%/ °C
Protein denaturation/coagulation	200-500	84%/ °C	47%/ °C
Enzyme reactions	100-200	28%/ °C	17%/ °C
Microbiological changes			
Microorganism growth	100-150	18%/ °C	na
Vegetative microorganism death	300-500	84%/ °C	47%/ °C
Spore death	250-350	53%/ °C	31%/ °C

* (The sensitivities are calculated using the highest activation energy in the range and at the temperatures quoted. Notice that they are temperature-dependent, which the Arrhenius equation requires.)

The sensitivities were calculated for the highest activation energy and at two temperatures: 40 °C and 120 °C. For example, taking general chemical reactions and the highest activation energy of 100 kJ/mol:

Sensitivity at 40 °C

E = 100,000 J/mol at 40 °C = 313 K, R = 8.314 Joules/mol. K

$E/RT^2 = 0.123$

$\exp\{E/RT^2\} = 1.130$

$= (1+ \text{sensitivity})$

Therefore sensitivity = 0.13 = 13% per degree C

Sensitivity at 100 °C

E = 100,000 J/mol at 120 °C = 393

$E/RT^2 = 0.078$

$\exp\{E/RT^2\} = 1.08$

$= (1+ \text{sensitivity})$

Therefore sensitivity = 0.08 = 8% per degree C

So both 13%/ °C and 8%/ °C sensitivities are seen to work out to the values shown in Table 2.IV. Notice that, according to the Arrhenius equation, sensitivities decrease with rising temperatures and conversely increase at lower ones for a fixed activation energy.

Looking generally, it is possible to view the whole realistic reaction spectrum available to the food processor. From this overview can come some understanding so far as temperature manipulation is concerned of:

- processing possibilities,
- degree of precision within which control has to be exercised if end results are to be defined and reproducible,
- degree of process control that is practicable, and
- accuracy that needs to be obtained from the instrumentation and the analysis of particular processes.

The results discussed in Example 2.4 are not from well-defined situations, but they are not untypical of exploratory work in processing.

Example 2.4: Temperature sensitivities in process development

You are faced with two processing problems in which working temperatures have shifted owing to circumstances beyond your control, so you have to accept them and make what time modifications you can to reach the same product results. One process is spore destruction and the other a hydrolysis.

For the spore destruction, the process is supposed to operate at 120 °C, but the temperature has fallen to 117 °C. By how much should you increase the standard processing time as a trial?

Contd..

Example 2.4 (contd)

From Table 2.IV, (1+ sensitivity) is 1.31 at 120 °C

The rate of spore destruction from 120 to 117 °C is predicted to decrease to $1/1.31^3 = 0.44$ times, i.e. 44% of its original value.

Therefore, processing times should be multiplied by 100/44 = 2.3.

For the low-temperature hydrolysis, the working temperature at 40 °C has risen by 2 °C. By how much should you decrease the former processing time to expect the same extent of hydrolysis?

From Table 2.IV, (1+ sensitivity) is 1.16 at 40 °C

The rate of hydrolysis, from 40 to 42 °C, is predicted to increase to $1/1.16^2 = 1.38$ times, i.e. 38% of its original value.

Therefore, processing times should be multiplied by 1/1.38 = 0.72.

If the standard time was 10 min, the run should be completed at the higher temperature in 10/1.38 = 7.2 min.

Sensitivities of process reaction rates to temperature can be used to design and control processes. They also give the product developer a feel for the experimental temperature ranges that should be tried in early experimentation, and also for control difficulties that might arise in the product line.

In a specific case, with more knowledge and tighter definition of the parameters, more precise process control could be exercised, and this is illustrated in Example 2.5. It should be noted that, even though the justification for the very large extrapolations implicit in the example may be questioned, relative changes calculated are rational and can be very helpful in guiding the process operator.

Example 2.5: Sensitivity of sterilisation processing

In a continuous fluid sterilising operation, you have included in your standard process calculations for sterilisation that your holding tube holds the product at 118 °C with a residence time of 7 min.

One morning, you discover from the product output that your fluid pump has unaccountably increased the flow rate by 30% and the only available way to correct in the short term and get usable product is by lifting the holding temperature. To what new temperature should you lift it?

You check out the sensitivity of the spore destruction, and find that it is 26%/°C at these temperatures. The activation energy for this spore death is 298kJ/ mol; the gas constant, is 8.314 joules/mol K; the temperature is 118 °C = 391K.

Contd..

Example 2.5 (contd)

$$\exp\{E/[R.T^2]\} = \exp\{298{,}000/[8.314.391.391]\} = 1.26$$

The flow rate has been increased by 30%, i.e 1/(1-0.3) = 1.43.
So you have to increase the rate of spore destruction by this factor.

Therefore, $1.26^{x} = 1.43$, which you find with your calculator = $1.26^{1.55}$.

So you need to increase the temperature by 1.55 °C, say 2 °C.

After this change, the process should then, assuming that heating and cooling contributions are substantially unaltered, slightly exceed the specification.

You do not need any tables, or a computer, although if you check them out you should find that they give the same answer. The added advantage of this 'first principles' approach is that the alarm bells of (technologically informed) common sense should ring because of the 'feel' that the 26%/°C gives for the sensitivity of the critical reaction, if results of calculations at any stage do not seem to be about right.

2.7 Reaction Rate/Temperature Relationship: Other Temperature Coefficients

There are several temperature coefficients used to characterise the effect of temperature on reaction rates. So far, only activation energies have been considered, together with a derived unit, sensitivity to temperature, that is convenient for quick calculations. There are more traditional measures that relate to these, two of them defined by the arbitrary symbols (**z**) and $\mathbf{Q_{10}}$, which are commonly used and will probably be familiar to many.

The small-case letter, z , is conventionally used to denote the experimentally found temperature increase necessary to multiply the rate of a reaction ten-fold. In our nomenclature,

z is defined by $\mathbf{10 = k_{T+Z} / k_T}$

The two *k*s are rates z degrees apart and z therefore is a temperature difference, and has the dimension of a temperature.

The symbol, Q_{10}*, is defined as the ratio of the reaction rate constants at temperatures 10 °C apart.* In our nomenclature,

$$\mathbf{Q_{10} = k_{T+10} / k_T}$$

The two *k*s are rates 10 degrees C apart and Q_{10}, being a ratio, has no dimensions.

The higher rate is in the numerator, so Q_{10} is always greater than 1. Q_{10} tends to not get much use these days although it was formerly common, especially in biochemistry.

Since z, Q_{10}, E, the activation energy and the sensitivity are all different ways of quantifying temperature coefficients of reaction rates, and they must all accurately fit to the same experimental findings, they must all lead to the same predictions. It follows that they must all be related to each other. Their full relationship is not straightforward, but, for practical purposes, useful approximate equations can be written to connect them:

$$\mathbf{E/RT^2 = 2.303 / z = \{\ln (Q_{10})\} / 10 = \ln (Q_1)}$$

where T is taken as the absolute temperature appropriate to the range of interest, often the mean in the range, and $\mathbf{Q_1}$ is a ratio of reaction rates, just as is Q_{10}, but with temperatures only 1 °C, rather than 10 °C, apart. So $Q_1 = (k_{T+1})/(k_T)$. As already discussed, Q_1 is closely related to the reaction sensitivity per degree, the numerical relationship being:

$$\mathbf{sensitivity = (Q_1 - 1) = \exp(E/RT^2) - 1}$$

so that, for a sensitivity of 13% = 13/100, $Q_1 = (1 + 13/100) = 1.13$.

An example of conversions of these measures is shown in Example 2.6, and some mathematical detail of the calculations is set out in Theory 2.3.

Example 2.6: Sucrose hydrolysis: conversions of temperature coefficients

The extent of hydrolysis with time was followed experimentally at several temperatures around 100 °C, and the reaction rate constants determined for each of the temperatures. These rate constants were plotted against the corresponding °C temperatures, and the slope of the plot showed the Sensitivity to be 10%/°C.

Now sensitivity = $(Q_1 - 1) = 10\% = 0.1$, and so $Q_1 = 1.10$

Using the relationships, $E/RT^2 = 2.303 / z = \{\ln (Q_{10})\} / 10 = \ln (Q_1)$, determine the value of Q_{10}, Z, and E.

Now	$\ln Q_1$	$= \ln 1.10 = 0.0953$
But	$\ln Q_{10}$	$= 10 \ln Q_1 = 0.953$, and so $Q_{10} = 2.6$
And	z	$= 2.303 / \ln Q_1 = 2.303 / 0.0953 = 24.2$ °C
And also	E	$= \ln Q_1 \, R \, T^2 = 0.0953 \times 8.314 \times 373 \times 373$
		$= 110{,}000$ J/mol

Of these measures, the most familiar to the food technologist will be z but the others relate quite straightforwardly.

Think break

Two reactions have z values of, respectively, 7 °C and 30 °C. At a working temperature of 70 °C (343 K), determine:

* the corresponding activation energies and
* the corresponding sensitivities.

For example, with a z value of 10 °C (= 18 °F) and at 70 °C, 343K

$$2.303/z = E/RT^2 = \ln Q_1,$$

$$2.303/10 = 0.2303 = E/\ 8.314\ (343)(343) = \ln Q_1$$

$$\text{so } E = 0.2303 \text{x} 8.314 \text{x}\ 343 \text{x} 343 = 225 \text{ kJ/mol},$$

$$\text{and } Q_1 = \exp(0.2303) = 1.26$$

$$= (1 + \text{sensitivity}),$$

so the sensitivity = 26%/°C

The detail of the relationships of these various temperature coefficients tends to be confusing and complicated. They have arisen from historical circumstances. For example, z arose about 80 years ago in the context of thermal death rates for spores in canning; it was convenient, it related to what was a useful temperature interval, and it became familiar. However, to work with it, particularly in the days before hand calculators, extensive tables were necessary. These were prepared and made available, but tables are always cumbersome and in factories they are messy. These days they have been incorporated into computer software that can work well, but software does not invoke much intuition in the technologist, or give a mental intimation when something has gone wrong.

Greater exploration of food processing in the wider field of reactions introduced activation energies more extensively, but they are also not very friendly to deal with either for calculations, or intuitively. This is why the temperature sensitivity, straightforward in concept and very easy to manipulate using a y^x key on a calculator, is used so freely in these present discussions. Some may find it interesting to work out the mathematics, and so these are briefly explored in Theory 2.3.

Theory 2.3: Temperature coefficients of reaction rate constants and their relationships

Writing the Arrhenius equation:

For temperature T_1

$$k_{T1} = A \exp\{-E/RT_1\}$$

and for temperature T_2

$$k_{T2} = A \exp\{-E/RT_2\}$$

Taking their ratios

$$k_{T1}/ k_{T2} = \exp\{-E/R[1/T_1 - 1/T_2]\} = \exp\{-[E/R][(T_2 - T_1)/(T_1T_2)]\} \;*$$

and putting $\varepsilon = T_2 - T_1$ then $T_1 = T_2 - \varepsilon$ and expanding

so $$k_{T1}/ k_{T2} = \exp\{-E\varepsilon /RT_2\}\{ 1- \varepsilon/T_2 + (\varepsilon/T_2)^2 - (\varepsilon/T_2)^3 + \ldots\ldots..(**)$$

Because ε/T is small (if ε is say < 20, and negative for $T_1 > T_2$) and remembering that E is large, say 100,000 + , and T is normally 300 +

Then by definition $Q_{10} \sim \exp\{10E/RT^2\}$ for $\varepsilon = 10$

or $\ln Q_{10} \sim 10\, E/RT^2$

and $Q_1 \sim \exp\{E/RT^2\}$ so $\ln Q_1 \sim E/RT^2$ for $\varepsilon = 1$

also by definition $10 \sim \exp\{z\, E/RT^2\}$ for $\varepsilon = z$

or $\ln 10 = 2.303 \sim z\, E/RT^2$

or $z \sim 2.303\, RT^2/E$

and so $E/RT^2 = (\ln Q_{10})/10 = 2.303 / z = \ln Q_1$ approximately

The series expansion (**) reveals the connection, and the reason for the fundamental differences and complications arising, between the Arrhenius approach, which includes the power terms, and the other (z, Qs), which can be seen to include only the first term in the bracket, 1. This follows from writing E/RT^2 as $\ln(Q_{10})/10$ or as 2.303/z. where T is a reference temperature.

For example, using the latter then from * above:

$$k_{T1}/ k_{T2} = \exp\{-2.303\, \varepsilon/z\} = 10^{-(T_2-T_1)/z}$$

because exp (2.303) = 10 and $\varepsilon = (T_2 - T_1)$.

This expression is found extensively in the standard food canning and thermobacteriology literature e.g. Stumbo (4).

For further exploration, there are complete books about these topics, such as Johnson *et al.* (3) for the Arrhenius equation in biological systems, and they are also discussed in general food processing technology books such as Singh & Heldman (5).

One remaining aspect of temperature sensitivity arises, and it is implicit in the approximations that are alluded to in Theory 2.3. When plotting temperatures against ln (reaction rate constants), the z plots give straight lines against the temperature, T, which can be either in °C or K. In contrast, the Arrhenius plots give straight line plots against the reciprocal of the absolute temperature, 1/T, in $(K)^{-1}$. This detail may seem confusing, but it is significant within the precision limits of most food processing work only when quite substantial temperature shifts (say 50 °C and up) are involved.

2.8 Reaction Rate/Concentration Relationships

The dependence of rate of reaction on concentration can generally be expressed by proportionality to some power of the concentration,

$$r = dC/dt = -kC^n$$

The value of *n* denotes the order of the reaction, n=0 being zero order, n=1 being first order. Whilst formally the exponent n in the general rate equation can have any value, the range is only of practical interest when needed for a fit or to help in understanding experimental observations in food processing. Two conditions, those of rates directly proportional to the concentration (n=1) and of rates that are constant (n=0), cover very many of the observed food processing situations.

Where they do not fit as closely as might be needed, then careful choice of averaged data can often improve fit. For example, rates calculated from slopes of concentration/time curves taken at a mid-point rather than an extreme, will often bring predictions into line with observations sufficiently closely to zero or first order for industrial purposes. Also, for the first 50% or so of any reaction, the orders are virtually indistinguishable.

2.8.1 First order reactions

Up to this point, first order reactions have been described, where the logarithmic increase or decrease in concentration is related linearly to time (see Fig. 2.2(b)). First order describes many useful reactions, and, even where it does not fit over the whole range, or exactly, it can still be a very useful approximation for part of the range.

Example 2.7: Sucrose hydrolysis: calculations of concentration changes with time

To see how concentration change is calculated, consider again the sucrose hydrolysis. It is known that its half-life at a particular temperature is 20 min. This means that starting at a concentration of 0.5 kg sucrose/litre,

- after 20 min it will have fallen to (0.50/2) = 0.25 kg/l, and
- after a further 40 min to $(0.25/2^{2)} = (0.25/4) = 0.063$ kg/l.

This can be extended to take into account any desired time by returning to the fundamental first-order equation:

$$\ln \{C_o/C\} = k\,t$$

and finding k from the fundamental first order constant/half-life relationship of

$$k = 0.693/t_{0.5}\,.$$

In this case, for a half-life of 20 min, $k = 0.693/20 = 0.035$ min^{-1}.
If the time to reach 0.20 kg/l is required, then:

$$\ln(0.5/\,0.20) = \ln(2.50) = 0.92$$
$$= k\,t$$
$$= 0.035\,t$$

$t = 0.92/0.035$, so that the required time is 26.3 min.

Or, if the sucrose concentration is wanted after 20 min,

$\ln(0.5/\mathrm{C}) = (0.035)\,(20) = 0.70$ and so
$C = 0.5/\exp(0.70) = (0.5)/2 = 0.25$ kg/l.

Fairly obviously, temperature and concentration changes can be combined, but the detail of this is best handled in a later section on process integration.

2.8.2 Zero order reactions

Another important class of food processing reactions is found experimentally to move at apparently steady rates, a constant amount decreasing or increasing rather than a constant fractional amount with time. These reaction rates can be seen in the overall scheme as proportional to concentration raised to the power of zero, and thus they are independent of concentration because $C^0 = 1$. They therefore move steadily at one constant rate and for them the calculations are rather straightforward, levels of reactants decreasing from their initial levels steadily with time. Unlike reactions of other orders, they continue to zero concentration within a finite time. The concentration increases or decreases with time linearly, as shown in the loss of ascorbic acid on storage of a multivitamin mix in Example 2.8.

Example 2.8: Ascorbic acid loss on multivitamin storage

In a classical investigation of loss of ascorbic acid in a multivitamin mix on storage at different temperatures, the concentrations of ascorbic acid at different times when stored at 50 °C were:

Time	(days)	10	20	30	40	50	60	70
Ascorbic acid	(mg/ml)	21	19	16	14	12	10	8

These data are shown in Fig. 2.4.

The actual figures in this case scarcely need analysis; the graph in Fig. 2.4 clearly shows the linear relationship with time, independent of concentration.

So this is a zero order reaction.

The rate k can be determined by taking:

(change in concentration)/(time taken) from the tabulated data or the graph trend line = (21 - 8)/(70 - 10) = 13/60 = 0.22 mg/ml/day.

Adapted from Garrett (6)

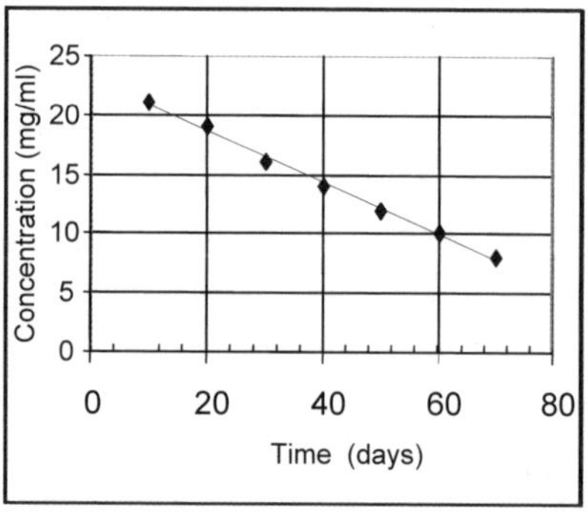

Data from Garrett (6)

Fig. 2.4. Ascorbic acid loss in a multivitamin mix on storage at 50 °C

Because it is the simplest form, it is always worthwhile to try zero order for any food processing situation to see whether it can be made to fit adequately for the purposes in hand.

Zero order reactions also conform to the Arrhenius equation in the relationship between temperature and reaction rate constant. In Example 2.9, data from Garrett (6) measuring rates of deterioration of vitamin C (ascorbic acid) in a multivitamin mix in storage experiments at different temperatures are given. They are plotted in Fig. 2.5 relating the logarithm of the reaction rate constant to the reciprocal of the absolute temperature. This is an Arrhenius plot showing that the data conform to the Arrhenius equation, and the graph is a straight line of slope ($-E/R$). Note that, in a zero order Arrhenius plot in Example 2.9, the reaction rate constant is in mg $ml^{-1}day^{-1}$, as compared with first order plot in Example 2.2, which is in min^{-1}. Reaction rate constant units are always for zero order concentration and time and for first-order time only.

Example 2.9: Loss of ascorbic acid on storage: Arrhenius plot

The reaction rate constants for an investigation of ascorbic acid deterioration in a multivitamin mix during storage at different temperatures were:

Temperature	(°C)	20	30	50	60	70
	(K)	293	303	323	333	343
1/T	(1/K)	3.41 $\times10^{-3}$	3.30 $\times10^{-3}$	3.10 $\times10^{-3}$	3.00 $\times10^{-3}$	2.92 $\times10^{-3}$
Reaction rate constant (k)						
	mg $ml^{-1}day^{-1}$	6.2×10^{-3}	2.4×10^{-2}	2.4×10^{-1}	6.6×10^{-1}	1.8×10^{-1}

k is plotted against 1/T in Fig. 2.5.

This conforms to an Arrhenius plot.

Adapted from Garrett (6)

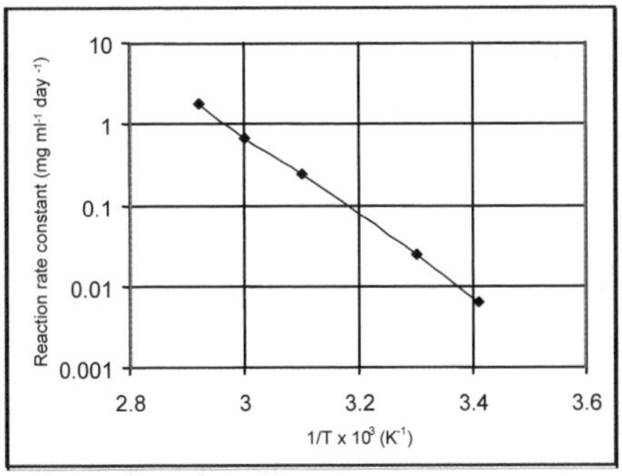

Data from Garrett (6)

Fig. 2.5. Ascorbic acid loss on storage: Arrhenius plot for reaction rate and temperature

2.8.3 Other rate/concentration relationships

There are some situations where other orders of the reaction are needed to secure sufficiently close fits, or where there may be some additional understanding of the reactions themselves resulting from more elaborate treatment of the data. A more practical reason for persisting with other orders is the examination of the general behaviour of reactions with order, and this is shown in Fig. 2.6, which shows the change of concentration with time. This displays the way in which ($1-C/C_o$) changes with relative time($t/t_{0.9}$); that is fractions of C_0 remaining as the concentration decreases from its initial value down to 10% of it (which can be called the 90% life) plotted against fractions of $t_{0.9}$. There are separate lines plotted for a number of different orders of reaction between 0 and 2. To cover the range more extensively, the time for a 90% change ($t_{0.9}$), rather than the 50% change in a half-life ($t_{0.5}$) has been chosen. The background to these curves is explained in Theory 2.4.

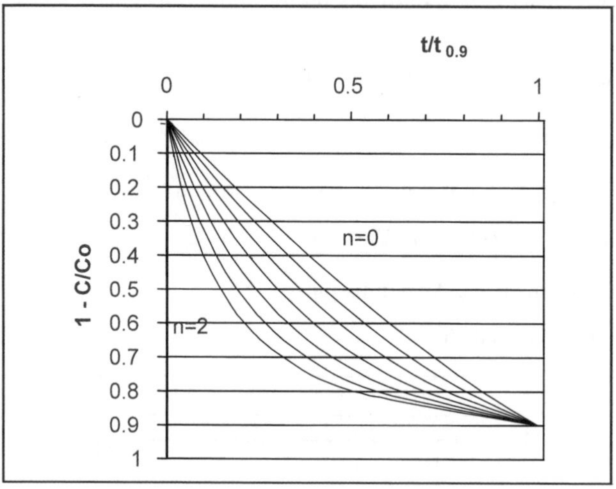

Fig. 2.6. Relative changes in concentrations with relative times for different values of reaction orders (n): 0, 0.25, 0.5, 0.75, 1, 1.5, 2.0, reference curves

Figure 2.6 shows quite clearly that higher order reactions progress more quickly, relatively, than do lower order ones in the initial stages. Then, to compensate, they move more slowly later as they get towards the common 90% point. Higher order reactions are relatively more sensitive to time in their early stages, compared with lower order reactions. And, of course, zero order reactions have constant progression with time, as would be expected. The graph also demonstrates that, with a suitable choice of a mean rate, taking a tangent rather than following the actual curve, any of the orders can be fitted closely by a straight line over the first 50% or so of the reaction. Thus, for many situations of practical relevance, zero order behaviour is a very reasonable approximation.

As an aside, but quite useful and sensitive in practical determination of order, it can be observed that the position of a given reaction on this graph is characteristic of its order. So, if an unknown reaction is plotted on such a graph, against a background of the characteristic lines for a number of selected orders, as in Fig. 2.6, the position of the line of the unknown reaction reveals its order (it fits over the curve appropriate to its own order).

These curves in Fig. 2.6 reveal clearly the considerable differences in the controllability of reactions of different orders as they move to completion, in terms of the extent (perhaps 50% or 90%) of the total change in their constituent components. Zero order moves steadily, at a constant rate, whereas, as the order increases, reactions move more quickly at first, then more slowly later and so the controllability of the reaction changes accordingly. As the order increases, their early rate of change with time is relatively much steeper for higher order reactions than for lower order reactions, so in their early stages higher order reactions are more difficult to control. If a critical reaction is of high order, and only proceeding to a few per cent, time sensitivity of concentration will be at its highest and control most difficult. Looking at the shapes of the curves also shows how the reactions of higher order depart much further from the simple approximation of uniform rate (graphically extrapolating by using a simple tangent at one point on the graph), and so order justifies being taken into account.

In foods that contain several simultaneously changing constituents, sensitivities may thus differ considerably between components, so some constituents have to be watched much more closely than others. When automatic controllers are substituted for human operators, this may have important implications for achieving consistency of quality. This discussion brings into consideration extents of reaction and their relative significance in different aspects of food processing.

The basis for the calculation of values of C/C_0 for different values of n against relative time $t/t_{0.9}$ is described in Theory 2.4.

Theory 2.4: Comparative progression of different orders: reference curves

The progression of reactions of different orders is shown by consideration of the reference curves.

From the general order integration equation

$$\{C^{1-n}-C_0^{1-n}\} = (n-1)\,k\,t$$

and if $t = t_{0.9}$ when the reaction is 90% complete, i.e. $C_{0.9}/C_0 = 0.1$

$$\{C^{1-n}-C_{0.9}^{1-n}\} = (n-1)\,k\,t_{0.9}$$

dividing one equation by the other to put $t/t_{0.9}$ in terms of the C's

$$\{C^{1-n}-C_0^{1-n}\}/\{C^{1-n}-C_{0.9}^{1-n}\} = t/t_{0.9}$$

then dividing through by C_0

$$\{(C/C_0)^{1-n}-1\}/\{(C/C_0)^{1-n}-0.1^{1-n}\} = t/t_{0.9}\,\text{(remembering that } C_{0.9}/C_0 = 0.1)$$

From this equation, values of 1 - C/C_0 can be calculated for different values of n against relative time $t/t_{0.9}$.

If (1 - C/C_0) is plotted against $t/t_{0.9}$ with order values n as a parameter at say 0, 0.25, 0.5, 0.75, 1.0, 1.5, 2.0, then the curves of Fig. 2.6 (called reference curves) result.

2.9 Relative Extents of Food Processing Reactions

The extent to which a critical food processing reaction proceeds is obviously of central significance to the processor. In our continuing example of processing, that of sucrose hydrolysis, reaction is desired to the extent of, say, 30% to 55% hydrolysed, that is in fractional terms C/C_0 running over a range from 0.7 to 0.45. This sort of figure covers a good number of food processing situations, and might be termed a medium range in the chemical sense. In other cases, substantially more complete reaction is required, for example in blanching, where 95-99% enzyme destruction might be necessary in order to sufficiently inhibit any undesired reactions that the enzymes might catalyse. Similar requirements might arise with nutritionally undesirable components, an example being protease inhibitors in soya beans. Chemical undesirables might also be in this same category, or they might be required to diminish even further. Rather loose nomenclature often speaks of removal when what is actually implied is removal down to below some hazardous, or sometimes analytically just detectable, level. This could be down to 0.1 or 0.01%, or lower, of initial levels depending on the particular hazard, or the extent of the 'undesirability'.

In other cases, the focus is on the products of a reaction, where in some instances even quite small concentrations may be significant. Examples are the off-flavours caused by some free fatty acids from fat hydrolysis, and the oxidation of fats creating rancid odours. In such cases, only a very small fractional change, say dropping the C/C_0 only to 0.99, that is 1% of some fats hydrolysed to fatty acids, may create detectable and unacceptable flavours and so be the 'processing' limit.

Looking at Fig. 2.6 it can be seen that, in the initial stages of a reaction, and up to 50% of the reaction completed, there is little practical difference between the different orders of reaction. For example, a straight line can be a reasonable approximation to them all. So the simplest constant rate calculations are adequate and these can cover all that matters in such applications. This may apply to storage reactions, which often appear to be zero order up to and beyond the point where sensory panels find the product unacceptable or below a necessary standard.

However, in another very important area, that of microbiology, the changes in concentrations can be quite dramatic. For microorganisms, the concentrations per unit volume or unit surface area can be interpreted as:

- most probable number (MPN) counts, or
- numbers of colony-forming units (cfu), or
- numbers of viable organisms.

For present purposes, these measures are all equally applicable and they are selected to suit particular circumstances. Increases in numbers of microorganisms, i.e. growth, can often be by multiples such as 10^4. In microbial death, measured experimental changes can be of the same order, whilst extrapolated changes (as in the conventional canning calculations to incorporate adequate safety) invoke multiples such as 10^{-12}. Putting aside for the moment any questions of justification for such large extrapolations, these differences in 'reaction extents' do not themselves provide problems in rate determinations. But they affect the apparent calculations when, as often happens in food processing, several simultaneous important reactions with differing extents occur at the same time in the same food product.

In summary:

- Reactant chemical composition in terms of C/C_0 changes from 1 at the beginning through to as little as 0.99 or 0.95 at a limit of acceptance in some critical flavour changes (and more often undesirable rather than desirable), through 0.5 to say 0.1 to 0.01 in component changes such as gelatinisation or protein denaturation in general processing.

- In bacterial or spore death, bacterial 'concentrations', taken in terms of, say, viable spores per can could move from, say, a measured 10^4 in one can down to an assumed 1 present in 10^8 cans, giving a total swing ratio of $10^{4+8} = 10^{12}$.
- In surface bacterial growth, bacterial 'concentrations' may move from a measured 10^3 per cm^2, to a measured and becoming obvious (smell, slime, colour) 10^7 per cm^2, which is a concentration ratio change from 1 at the beginning to 10^4 ($10^7/10^3$).

Case study 1: Retention of vitamin A in liver processing

This case study shows how some laboratory research can be used as a basis for designing a heating process.

For foods that are significant sources of essential nutrients in the diet, it is obviously important that as much as possible of the nutrient content, for example of vitamins, is retained for the consumer if the food is processed. This was the subject of a study on the vitamin A content in a processed liver product intended for an infant food by Wilkinson *et al.* (7). Liver is a rich source of this vitamin, so that, if the liver is heat processed to make a baby food, it is desirable to retain as much as possible commensurate with meeting whatever other processing stipulations may be dictated, such as, for example, ensuring microbiological safety.

The experimental laboratory study of the retention of vitamin A in liver during heat processing showed systematic behaviour. The loss of vitamin A on heat processing conformed closely to the first order pattern (Fig. 2.7a).

At 126.7 °C (399.7 K) the vitamin A concentration as a function of heating times, was:

Time (min)	0	10	16	22	28	34
Concentration (μg/g)	271	109	58	30.5	18	10

The plot of log (concentration) against time in Fig. 2.7a gave a good straight line, so fitting first order.

Contd..

Case study 1 (contd)

The first order rate-constants and corresponding temperatures were:

Temperature	(°C)	102.9	111	118.3	122.1	126.7
	(K)	375.9	384	391.3	395.1	399.7
$1/T \times 10^{-3}$	(K)	2.66	2.60	2.56	2.53	2.50
Reaction rate constant *k* $(s^{-1}) \times 10^5$		17.9	38.6	68.0	96.1	162.3

The relationship between 1/T and the reaction rate constant is shown in Fig. 2.7b. and the effect of temperature is seen to follow the Arrhenius relationship.

From the slope of the line in Fig. 2.7b, the activation energy was calculated as 112kJ/mol. This activation energy translates into a sensitivity of the reaction at 118 °C of around 9%/°C.

As an example, work out the residual concentration of vitamin A initially 300 μg/g, after processing for 20 min at 120 °C. We know that $k_{118..3} = 68 \times 10^{-5}$, and with the sensitivity as 1.09, then $k_{120} = 68 \times 10^{-5} (1.09)^{(120-118.3)} = 78.2 \times 10^{-5}\ s^{-1}$. And so, for 20 min at 120 °C, $kt = 20 \times 60 \times 78.2 \times 10^{-5} = 0.938 = -\ln C/C_0$, and therefore $C/C_0 = 0.39$ and so the initial concentration of vitamin A would be reduced from 300 to 300 x 0.39 = 117 μg/g.

This is about a 60% reduction and would obviously be of importance if the purée were effectively the only source in the diet of an infant and if the quantity provided daily had been prescribed on the basis of unprocessed liver. Reducing the processing time to 10 min, if this were permissible, would reduce the loss to 37%, and further reducing the processing temperature to 115 °C, the loss would be only around 26%. So it can be seen that the technique provides the processor and the formulator with useful information, in which various processing strategies can be compared. This illustrates the application of reaction technology in producing data applicable to industrial production, in this case of baby food, providing necessary information for the specification of the processing conditions so that an infant diet ingredient could contain sufficient of an important constituent.

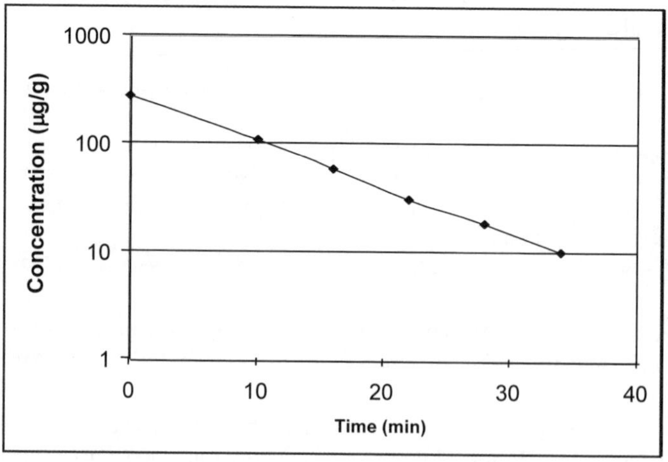

Data from Wilkinson *et al.* (7)

Fig. 2.7a. Vitamin A loss on processing: change of concentration with time at 126.7 °C

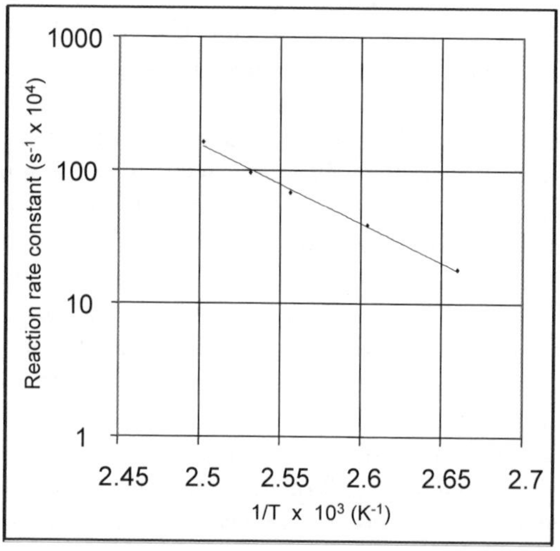

Data from Wilkinson *et al.* (7)

Fig. 2.7b. Vitamin A loss on processing: change with temperature: Arrhenius plot

Case study 2: Yellowing of whey protein coating on storage

This case study illustrates the use of reaction technology in storage tests. It is an example of the use of a physical method, reflectance measurement of colour, as the measure of change.

Panned confectionery products can be coated to give them gloss and to protect their exterior, and whey protein concentrates can be used in these coatings. However, on storage, the coatings gradually change from almost colourless to deeper shades of yellow, and this has been studied as a function of time, as reported by Trezza and Krochta (8). The yellowing was measured by an arbitrary but standard (ASTM) reflectance method. The increase in yellowing with time at 23 °C was:

Storage time (months)	0.09	0.27	0.69	1.92	3.1	4.33	5.0	6.48	7.96	9.76
Yellowness Index	9.7	9.8	10.0	9.8	11.1	11.8	12.7	13.3	14.8	17.0

These results have been plotted on Fig. 2.8 and show a reasonable straight line over a considerable period of storage (10 months).

Thus the time before some critical degree of discolouration is reached, say 15 units, could be read from the graph.

The rate of increase of the yellow colour can be determined either crudely from the extreme values:

$$(17.0 - 9.7) / (9.76 - 0.09) = 0.75 \text{ units per month}$$

or more accurately from the trend line of the graph, established most readily by computer program, which is shown as the trend line equation on the graph. In the case of the 23 °C results, where C_0=9, this is (rounding off):

$$\text{Yellowing} = 0.7\ (\text{months}) + 9$$

Also plotted on the graph are storage results for temperatures of 40 °C and 55 °C, which plot to give straight lines.

When the slopes of these lines were plotted against the reciprocals of the absolute temperatures in an Arrhenius plot, an activation energy of 95 kJ/mol was obtained and this can be used, for example to determine storage lives under warehousing temperatures.

Contd..

<u>Case study 2 (contd)</u>

The authors commented that, with the higher temperatures, there was deviation from straight line behaviour after 4.5 weeks, which is why results for longer periods are not shown. However, if, say, 15 units were the maximum tolerable, then the zero order behaviour could be used for the predictions over the whole temperature range reported. The equation indicates that the 'storage life' at 23 °C to a maximum yellowness of 15 should then be about 8.5 months. The other lines with their equations would indicate lives at their respective temperatures, and the activation energy with the Arrhenius equation would fill in for other storage conditions as needed.

The study demonstrates that arbitrary units can be used if convenient, and particularly when there are no fundamental units available. Units chosen do need to be consistent and reproducible, and there are obvious advantages if they fit standard specifications, as do these, and can therefore be reconciled with measurements in other situations.

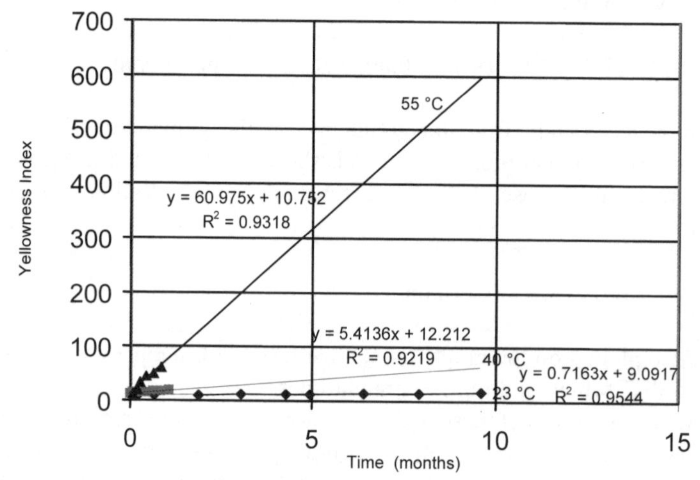

Data from Trezza & Krochta (8)

Fig. 2.8. Yellowing of whey protein on storage

2.10 Practicalities

In practical terms, how can this approach, rate equations and temperature coefficients, be incorporated usefully into industrial food processing? The first step is to obtain adequate data. It may be that suitable results can be obtained from the food research literature and data tables; there is often a surprising amount available. In larger companies, there may be laboratory and factory trial results in the files. Even in small companies there will usually be experience, and this can at least give ideas as to rates and even to temperature sensitivities. Otherwise, experimental evidence is needed and this means measuring the rates at which the desired processes actually progress. The steps in building a reaction technology approach to process control and development are shown in Fig. 2.9.

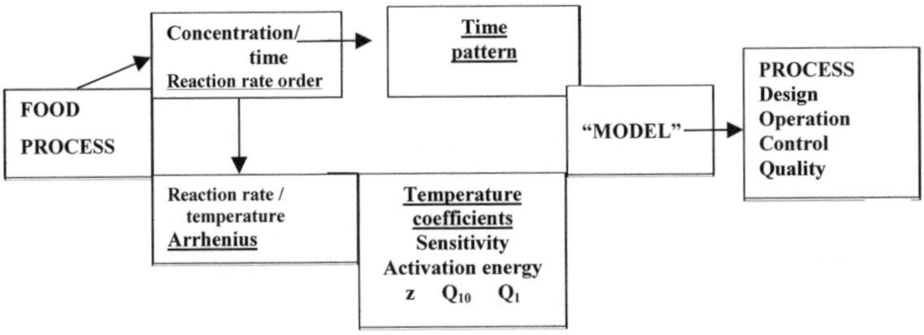

Fig. 2.9. Overview of reaction technology approach

2.10.1 Studying change in concentration with time

'Suitable' readings of the relevant measures at 'suitable times' are made as the reactions progress. 'Suitable' is something that has to be found by trial and error, but common sense and experience produce it quite quickly. The measures have to be related to the necessary controls during processing and the specified product attributes. Refinement thereafter is just natural progression. Experimentation can be done by following one concentration variable at a time, keeping other ingredient concentrations and the temperature constant. The alternative is to use systematic experimental designs and to study several variables together over the practical range, taking limit values and sorting out the effects of the different variables statistically. The traditional approach is simpler but sometimes it is less efficient, not practicable, or unduly cumbersome.

What emerges is a set of times and corresponding concentrations. Initially, these can be plotted linearly, and if a straight line fits then the reaction is zero order. For the other orders, it is easiest to set up the corresponding version of the general equations in linear form, so that straight lines plots result when you have got it right. For example, this means plotting the logarithm {ln C} against time for first order, and the reciprocal of the concentration {1/C} against time for second order, as illustrated in Table 2.V.

TABLE 2.V
Straight line plots for reaction orders

Plot experimental data using X axis and Y axis to produce a straight line for different orders:

Order	X axis	Y axis	Slope Y/X
0 order	time	C	- k
1 order	time	ln C/C_0	k
2 order	time	$(1/C - 1/C_0)$	k

where 'C' is typically concentration, but may be some other convenient, consistent measure, and k is the reaction rate constant.

The reaction rate constants (k) can be determined from the slopes of the graphs.

2.10.2 Studying rate of reaction/temperature relationships

So we end up with an order and a rate constant. This has then to be repeated over a range of temperatures. From the resulting set of rate constants and corresponding temperatures via an Arrhenius or similar plot, an activation energy, a z value, or a temperature sensitivity can be calculated, according to preference. Remember that the natural logarithm of the reaction rate constants is plotted against 1/T (in degrees K) for activation energies, and for all the others against just T (which can be in °C or K as this just moves the zero), as summarised in Table 2.VI.

TABLE 2.VI
Straight line plots for temperature coefficients

Plot experimental data to obtain temperature coefficients of reaction rates:

	X axis	Y axis	Slope Y/X
Arrhenius(1/T in exponent)	1/T (K)	ln (rate)	-E/R
z (linear T in exponent)	T (°C or K)	ln (rate)	-2.303/z
Q (linear T in exponent)	T (°C or K)	ln(rate)	$-(\ln Q_{10})/10$

If you cannot distinguish the orders within the precision of the results, and yet this precision is adequate for demands of assuring the final product, then order is probably not critical. For very many practical purposes zero or first order, if carefully fitted and maybe with a little extra care in selection of average rather than extreme values, sufficiently good predictive systems for industrial use can be produced. The worked examples and figures illustrate some of the important points. When predictions involve major extrapolation, always keep in mind that even quite minor poorness of fit of the equations can build up into substantial discrepancies between predicted and observed outcomes.

2.10.3 Studying temperature coefficients

What is often much more reliable than trying to predict an output from scratch is where the output under one set of conditions is already known and you have only to predict how it changes after a known change in conditions. Then the resulting movement in the outcome can be predicted through the sensitivities to the change. In other words, when process variabilities, as they always do, push the process away from the correct settings, then these methods will help predict how to move quickly and accurately back to where it should be.

You may have found sensitivities easy and convenient, but, if not, of course all the other measures work as they must if they are reliably based on real data. If the change in more than one constituent is important, then you have to go through the procedures for each. Take comfort from the fact that, if such constituents are in fact important, then they justify attention.

The resulting data and the framework into which they have now been fitted can be used in predictive calculations. They can also be used in 'what if' scenarios, in examining and specifying artefacts like instrumentation and control valves and heat transfer surfaces so that they are responsive to process disturbances with the speed and the precision needed to ensure the product quality required, in determining suitable 'ambient' reaction conditions such as water activity, pH and gas pressures, and in preparing charts and other aids for operating personnel.

As a very simple example, if operating a process with a critical reaction having a sensitivity of around 25%/°C, then a thermometer that is unable to discriminate reliably to better than 1 °C can present problems for control.

2.10.4 Time patterns

The changes with time, and the changes of times with temperature are very important, and are discussed in detail in Chapter 3.

From all this information, a 'model' can be built for the process. Very often the model is in reality a set of quantitative relationships with which reliable predictions of component behaviour can be made. It may be simple or complex! Remember that, with all of these methods, if the analysis is reliable then interpolations will be just as reliable, but extrapolations should always be treated with caution and the greater the extrapolation the greater the caution needed. However, much of the point of the analysis is to be able to extrapolate, and so familiarity with the data and with the analytical framework and its basis, and experimental checks, using critical cases, will build both confidence and skills.

2.11 References

1. International Critical Tables. New York. McGraw-Hill, 1927.
2. Levenspiel O. *Chemical Reaction Engineering 3rd Edn.* New York. Wiley, 1996.

3. Johnson F.H., Eyring H., Stover B.J. *Theory of Rate Processes in Biology and Medicine*. New York. Wiley, 1974.

4. Stumbo C.R. *Thermobacteriology, 2nd Edn*. New York. Academic, 1973.

5. Singh R.P., Heldman D.R. *Introduction to Food Engineering, 2nd Edn*. San Diego. Academic, 1993.

6. Garrett E.R. Prediction of stability in pharmaceutical preparations II. Vitamin stability in liquid multivitamin preparations. *Journal of the American Pharmaceutical Association*, 1956, 45, 171.

7. Wilkinson S.A, Earle M.D., Cleland A.C. Kinetics of vitamin A degradation in beef liver puree on heat processing. *Journal of Food Science*, 1981, 46, 32.

8. Trezza T.A., Krochta J.M. Color stability of edible coatings during prolonged storage. *Journal of Food Science*, 2000, 65, 1166.

3. PROCESSING OUTCOMES

3.1 Introduction

In the previous chapter, rates of change in foods under constant processing conditions were introduced, but many food processes have changing conditions. What matters to the technologist controlling processing is the overall consequence of the change in the materials during processing – the outcome, the final product. The questions to answer are: what can be obtained from the raw materials, and what processing conditions produce this product? From a technical viewpoint, this means summing all of the step changes that occur over the total process time to give the final product attributes. Then the process conditions in each step can be manipulated to best advantage by the operators to give the optimum final product attributes. The extent of each reaction is controlled and thus the specified levels of the critical and important product attributes can be achieved. In this chapter, the processing conditions of time and temperature are studied. There are other conditions, such as relative humidity, pressure, packaging, and surrounding atmosphere, that also affect the reaction rates, and these are discussed later.

3.2 Steady Conditions of Time and Temperature

Early in the exploration of systematic food technology, questions were asked about the extents to which necessary processing times could be altered by varying processing conditions. One important practical application was heating under varied pressure, in which the working temperatures were dropped by processing in a closed container under a vacuum, or lifted by processing under increased pressure (1). Another practical application explored very extensively (2) because of its great consumer significance was killing microorganisms in heat sterilisation or reducing microorganism numbers in foods by heat pasteurisation. It was found, not surprisingly, that, as temperatures were increased, times required for a given process were decreased, and vice-versa. But perhaps more surprising were the early findings that these changes were systematic and consistent, and then later on that the changes could be fitted into the patterns of reaction technology and were therefore predictable.

It is first necessary to build the general patterns of processing reactions. This will enable the logic to be demonstrated, but it is more important to build up a logical platform on which future extensions and applications can be erected. The immediate need is to determine how the necessary times relate to the working temperatures for a defined change in the food achieved through processing. Careful inspection of the equations developed in Chapter 2 show that this question has already been answered.

For a first order reaction, such as acid hydrolysis of sucrose, it was shown that:

$$t = (-1/k_T)\ ln\ (C/C_0)$$

or $$\mathbf{-k_T\, t = ln\ (C/C_0)}$$

This tells us that the product of k and t is constant, for a defined extent of processing in which the targeted concentration goes from initial value C_0 to a final value C. Now k changes by changing the temperature T; therefore, the required time t for the same extent of processing will also have to change with temperature change, and more significantly, change so as to keep the product kt constant.

There is nothing that is specific to first order except the form of $ln\ (C/C_0)$ arising from the integration. Therefore, the constancy of (kt) is equally true for a reaction of any order moving at a constant temperature between a particular initial value C_0 and a particular final value C. This is very useful. It means that a graph of rate of change against temperature can be re-figured to produce a graph of reaction time against temperature. This can be called an outcome/time-temperature (OTT) chart. The theory for the OTT chart is shown in Theory 3.1.

Theory 3.1: Basis for the OTT chart – integration of general rate equations

For the typical process looked at generally:

$$\text{rate (r)} = dC/dt = -\ k\ (\mathrm{T})\ f(C)$$

where $f(C)$ is a function of concentration (for a first order reaction $f(C)$ is kC) and writing k(T) draws attention to k being a function of temperature (T), whereas k_T denotes k at a particular temperature level T.

Contd..

Theory 3.1 (contd)

Following the Arrhenius equation, $k = k(T) = A \exp(-E/RT)$

and so $dC/f(C) = -k(T)\,dt = -A \exp(-E/RT)\,dt$

in which the left-hand side (LHS) is some function of *C,* which can be integrated algebraically or numerically,
and the right-hand side (RHS) contains time and on integration gives the time needed for the change in the relevant food ingredient from C_O to *C.*

$$\int_{Co}^{c} dC/C = -\int_{0}^{t} k\,dt \qquad = -kt \quad \text{if k is constant}$$

$= \ln C/C_0$ for a first order reaction
$= (1/C - 1/C_0)$ for a second order reaction, and so on.

If temperature varies with time, along some experimental temperature/ time curve, T = T(t) then k is not constant over the time and so:

$$\int_{Co}^{C} dC/C = -\int_{0}^{t} k\{T(t)\}\,dt$$

The LHS is unchanged but the RHS has to be integrated, either analytically if T(t), expressing T as a function of time (t) can be suitably expressed (27), or numerically/graphically if it cannot.

For a defined change between the concentrations C_0 and C, and irrespective of the form that the relationship takes (zero, first, ...order), the LHS is constant between any particular limits C_0,C, and therefore the RHS must total to that same constant sum.
If measurements are made of k at different temperatures and the data are found to fit the Arrhenius equation, on plotting ln k against 1/T where T is measured in degrees K, a straight line with slope –E/R results. This is the traditional Arrhenius plot.

For a particular concentration extent of a first order reaction, the product kt is constant and this can be put = K. Then:

$\ln k$ is equal to $\ln (K/t) = \ln (K) - \ln (t)$

and so a plot of ln (t) against 1/T will also produce a straight line with slope E/R but displaced vertically by ln(K). This result is equally true for orders other than one. This is the basis of the OTT chart.

The Arrhenius graph can thus be converted to the outcome/time-temperature (OTT) chart, remembering that each line on this chart connects conditions for one specific extent only. For the hydrolysis of sucrose, the Arrhenius plot of reaction rate constant against the reciprocal of T (1/T) (Fig. 2.3) can be converted to a similar plot of time against the reciprocal of temperature in degrees K (Fig. 3.1(a)) or into an OTT plot of time against degrees C (Fig. 3.1(b)). The X-axis in the OTT chart is a linear scale in degrees C.

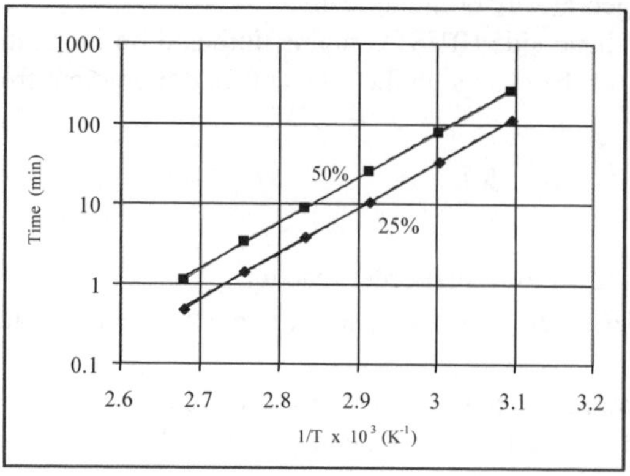

Fig. 3.1(a). Sucrose hydrolysis – "Arrhenius" plot for time

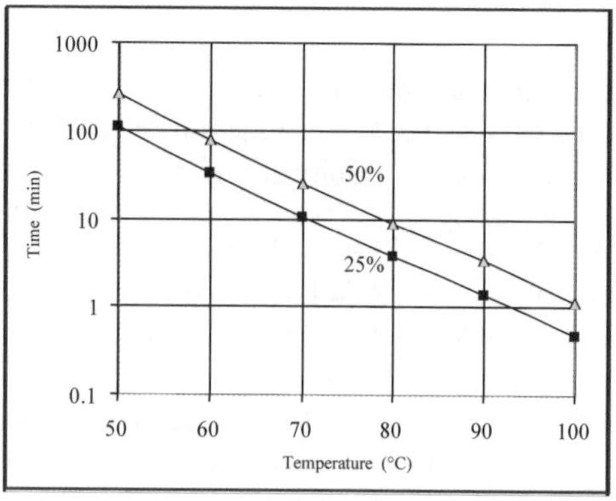

Data from International Critical Tables (3)

Fig. 3.1(b). Sucrose hydrolysis – OTT chart

Note that the slopes of the lines are opposite because of temperature and reciprocal temperature. Looking carefully at these graphs will show that, whereas the lines in Fig. 3.1(a) with reciprocal temperature in degrees K are straight, those in Fig. 3.1(b) with linear temperature in degrees C are slightly curved. The reciprocal temperature correlation fits the data slightly better than the straight temperature, although the discrepancy is small and can be ignored within the tolerances of most processing specifications. Although these are not equivalent mathematically, they are approximately so, and more nearly so as the total temperature interval gets smaller (say 10 or 20 °C). For intervals of 10-20 °C, either can be used for calculations with accuracy as good as that of the available data. Experimental data for most reactions for small temperature intervals show that a plot of both 1/T and also degrees C against the logarithm of the reaction time to reach a particular outcome follow a straight line. For larger intervals, the 1/T plot generally fits better, so should be preferred unless comprehensive data show otherwise.

The OTT presentation is a very powerful and useful tool, and one that has a number of applications in food process technology including on the factory floor. The OTT chart can be drawn for various outcomes such as 25% reacted, 50% reacted, and so on, with a separate line for each outcome as shown in Fig. 3.1(b). These clearly show the effect of time and temperature on the outcome. They can readily be extended to give data for any desired degree of hydrolysis that may be appropriate to a particular processing need. From such Arrhenius/time, and OTT charts, time and temperature combinations can be chosen for specified product attributes.

Think break

- Consider and reflect on the outcome/time-temperature (OTT) chart and the possible uses in food processes with which you are familiar.
- If in a particular sugar boiling, the aim is to reach 50% hydrolysis, how long would this take at 50, 80 and 100 °C?
- Explain how the sucrose hydrolysis OTT chart could be directly applied by an operator boiling jam in vacuum pans where the process steam supply variation can cause a temperature variation of 10 °C.

When the conditions of processing are constant, and in our case this implies at constant temperature, the extent of the reaction is obtained from the rate per unit time multiplied by the time interval over which the process continues. For this use, the symbol ∇ is sometimes used for convenience; therefore $\nabla = \boldsymbol{kt}$.

For a first order process, $\nabla = -\ln C/C_0$
Note ∇ is dimensionless

It is sometimes expressed in terms of decimal (D) reductions. For example:

- when $C = 0.01C_0 = C_0/10^2$, this is called a 10^2 or a 2D reduction (and $\nabla = -\ln 10^{-2} = -2.303 \log 10^{-2} = 4.61$);

- when $C = C_0/10^3$ this is called a 10^3 or a 3D reduction (and $\nabla = -\ln 10^{-3} = 6.91$), and so on.

This nomenclature is particularly convenient for processing changes in bacterial numbers, or counts, and these may reach 10D or more.

3.3 Variable Conditions of Time and Temperature

In a process, conditions may vary as the process goes along; for example, on heating a starch solution in a jacketed pan, the temperature will rise gradually from ambient to boiling, i.e. 100 °C. There can also be variations in temperature with space – for example, in a can, where parts heat faster depending on the consistency of the food material and the type of steriliser (static or continuous).

3.3.1 Sequential changes in temperature with time

Where temperatures are changing during the process, the times can be converted to the time at a reference temperature. Then the times can be added to give the total time of processing at the reference temperature.

Temperatures can vary over quite a wide range, but for food processing there are some *reference temperatures* on which attention can readily be focused.

- Boiling point of water at atmospheric pressure, 100 °C. Water is a universal constituent of food; often it is the major component, and at altitudes not too far from sea level water boils at close to 100 °C over the normal ranges of atmospheric pressure variation.
- Two atmospheres absolute pressure, under which pressure water boils at 121 °C (250 °F). This is based on pressure processing, and canning in particular, where an early and convenient 'pressure cooking' level was at around two atmospheres absolute pressure (or one atmosphere above atmospheric pressure). Under this pressure, water boils at a temperature of close to 250 °F, so this, and nowadays its Celsius equivalent of 121.1 °C, which can be rounded to 121 °C, forms a good reference temperature.
- Frozen storage, -18 °C (0 °F). Below-freezing temperatures for frozen storage have never really needed a reference temperature. A quite widely encountered reference temperature has emerged of 0 °F or -18 °C, possibly for no better reason than that it is a conveniently low level for much of frozen food storage and is such a nice round number on the Fahrenheit scale.
- Chilled storage. There is no fixed reference temperature, but many chillers operate around 4 °C. With superior control, they can be operated down to just above food freezing temperatures of around -1 °C. Reference temperatures of 0 °C and 5 °C have been used.

> *Think break*
> For two food processes – concentrating milk in an evaporator, and chilling of fruit after harvest:
> * What are the important temperatures in the two processes?
> * What is the temperature history in each process – how does temperature change with time?
> * What reference temperatures could be useful in standardising the temperature criteria?

So for a constant temperature process, $kt = \nabla$ and is constant for a given level of processing. If k_{ref} is the value of k at the reference temperature, then a value of t_{ref} can be calculated:

$$\mathbf{t_{ref} = \nabla / k_{ref} = k_T . t_T / k_{ref} = (k_T / k_{ref})\, t_T}$$

where k_T is the reaction rate constant and t_T is the processing time during a certain step of the process at a constant temperature, T.

F values are encountered in canning and sterilisation, where the reference temperature is 121 °C, and t_{121} has been given a special symbol of its own, F. More strictly speaking, any relativities involve not only a reference temperature but also the sensitivity of the rate of reaction to temperature. Then either activation energy (E) or z or Q_{10} values should be designated for the reference data.

For example, in sterilisation, F_x^y , where x and y are the appropriate z and temperature reference values, traditionally F_{10}^{121} is given the symbol F_0. So a whole range of equivalent processes arises, such that the extent of sterilisation at any temperature can be expressed in terms of an equivalent time of processing at the reference temperature by working out its F_0 value, remembering that this refers only to the standard 'canning' system with z value of 10 °C.

C values have been defined, particularly for cooking where the reference temperature is the temperature of water boiling under atmospheric pressure. So the time of heat processing at 100 °C, t_{100}, can be called the 'cook time' and given its own symbol "C". Equivalent cook times for processing can be worked out at 100 °C. Examples are cooking at much elevated locations, where the atmospheric pressure, and therefore the boiling temperature of water, are substantially lower than at sea level, cooking at some non-boiling temperature, and comparisons between cooking at different temperatures are sought. Strictly, a cook time should define the reference temperature (100 °C) and the z value. In Example 3.1, values of F_0 and C for sucrose hydrolysis are calculated.

Example 3.1: Determination of F_0 and C values

(1) Find the F_0 value for a process taking 1 min at 128 °C.

From the basic relationship, kt = constant

Therefore $k_{T1}.t_{T1} = k_{T2}.t_{T2}$ and so $t_{T1} = t_{T2}(k_{T2}/k_{T1})$

But "z theory" holds that $k_{T2}/k_{T1} = 10^{(T2-T1)/z}$

And for standard F_0 calculations, z = 10 °C and for T_1 = 121 °C, $t_{T1} = F_0$

So $t_{T1} = F_0 = t_{T2} \times 10^{(T2-T1)/10} = t_{T2} \times 10^{(T2-121)/10}$

Now for the example T_2 = 128 °C and t_{T2} = 1 min

So $t_{121} = F_0 = 1 \times 10^{(128-121)/10} = 1 \times 10^{0..3} = 1 \times 5.0 = 5.0$min*

(* This calculation has been chosen to illustrate the asterisked term in Example 3.3)

Compare by "Arrhenius theory" for z = 10 at 121 °C, then E = 297 kJ/mol and T_1 = 394K, T_2 = 401K

$$(k_{T2}/k_{T1}) = \exp\{ E/R\,(T_2 - T_1)/T_1 T_2\}$$

$$= \exp\{297{,}000/8.314\ (401-394)/394.401\}$$

$$= \exp 1.58 = 4.9$$

and so $t_{121} = F_0 = t_{T2}(k_{T2}/k_{T1}) = 1 \times 4.9 = 4.9$ min

(2) Find the C value for a cooking process of 10 min at 95 °C.

To show how equivalent times other than F_0 can be calculated, find the C value of a cooking process of 10 min at a temperature of 95 °C. For this, the reference temperature is 100 °C and the z value 24.7.

So $C = t_{100} = 10 \times 10^{(95-100)/24.7}$

$$= 10 \times 10^{-0.2} = 10 \times 0.63 = 6.3 \text{ min}$$

and writing this formally $C^{100}_{24.7}$ = 6.3 min.

This could also be worked out by Arrhenius relationships.

Notice that C here should have been designated $C^{100}_{24.7}$ (100 being the reference temperature in degrees C and 24.7 the chosen z value for which C_0 is sometimes used.).

> *Think break*
> Reflect on the following time/temperature relationship as the measures of temperature sensitivity for reactions in foods:
> * contrast OTT charts using degrees C and Arrhenius time plots using 1/T as the x-axis.
> * contrast activation energies with z values
> * suggest processes in which it would be useful to use each method.

3.3.2 Space and time/temperature

The relationship of temperature variations to reaction rate constant variations merits further attention because it introduces important areas that complicate not only calculations of reactions in food processing but also the outcomes in the final food product.

In a steam-heated experimental static retort, temperature measurements were taken during a complete process cycle at three different positions along the centreline of a can, held vertically and stationary and filled with water (4). The temperatures were carefully measured, using fine thermocouples mounted in a sealed commercial-sized can (7.8 cm x 11.6 cm) with a liquid fill and air headspace, and were recorded at intervals during the process cycle. A group of these temperatures is shown graphically in Fig. 3.2.

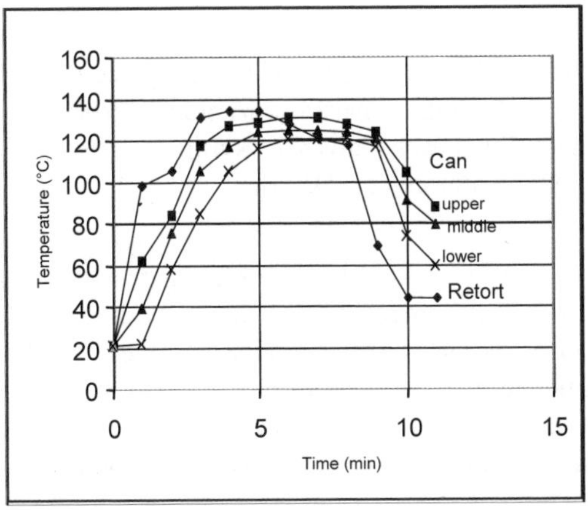

Data from Packer (4)

Fig. 3.2. Temperature profiles in a can of water during static retorting

The time/temperature graphs are plotted from experimental data for three different points: one at the geometric centre (middle), and two on the long centre line at the upper and lower third-points. This shows the considerable variation of the temperatures at different times during the processing, and therefore of the reaction rates. Note, in this case, the temperatures of the lowest point in the can rose more slowly than those of the middle point; in the case of more solid packs, the centre could be expected to rise the slowest.

F_O values were calculated at these three points in the can for spore destruction (see Example 3.3). Such data can be incorporated in calculations working towards a volume-based average, in other words adding the various component concentrations weighted in proportions to their relative volumes. Then the final number of organisms present in the total volume of the food can be obtained by multiplying that volume by the average concentration of the organisms.

Complications that can arise include the following:

- Inadequate mixing – for example in semi-solid foods in static retorts – can cause variation in composition and counts.

- In many foods – meat for example – the initial muscle tissue is sterile and the only contamination may be on the surface, so that a volume average is not meaningful. Mincing or cutting the meat will alter this situation, but may still produce great variations in counts per unit weight throughout a volume.

- In processing liquid foods – say, by pumping them through a pipe heat exchanger, the liquid velocity distribution is not uniform, often with an average velocity only about half that at the centreline. This means that an element on the centreline spends only half as long in the process as an average element, let alone one in the slowest moving regions towards the wall of the pipe.

- In food that is non-homogeneous, with aggregates of one component in a continuous mass of another, there is variation in process extent between the liquid and the solid. Peach halves in syrup are a simple example; casein micelles in milk are a more complex one.

These problems can be tackled by considering residence times at particular positions in the food being processed, looking at both the temperatures and the process time of all of the parts of the food where these are different. Averages, extremes, and particular elements can all be critical. The analysis is complex and sometimes can be safely ignored, but the processor has to be alert and aware that it may be important and that it can, if necessary, be explored in detail.

Think break

Consider the variations in temperature at different levels in a 2 kg beef roast of 10 cm minimum dimension, heated in an oven at 170 °C to an internal temperature of 65 °C in the centre.

* Estimate the initial and final temperatures at the different parts of the food structure.
* Sketch the possible final temperatures from the outside to the inside of the roast.
* What are the required product attributes at the different levels of the meat ?
* Identify the important reactions occurring in the meat that affect the final product attributes.
* Identify the reaction rate constants for these reactions or predict comparative rates for the reactions, e.g. from Table 2.IV.
* Reflect on the consequent relative extents of processing that might occur at different points in the foods.
* Reflect on how some foods, such as roast meat, depend on these temperature differentials to provide structures such as the crust and the crackling.
* What would be the impact of different cooking times and oven temperatures on the product outcomes?

3.4 Microbiological Outcomes from Process Reactions

Microbial constituents of foods can be treated in many respects as concentrations, that is by considering the number of viable organisms, normally measured as colony-forming units (cfu) in unit volume, although sometimes the number per unit surface area is more pertinent. These numbers change, by growth, by decay, or by processing to deliberately reduce or increase them.

3.4.1 Microbial growth

Microorganisms increase by division, and it has been found experimentally that, after initiation, called the lag phase, with no or little growth in cell numbers, the divisions occur at a fairly constant rate and give rise to logarithmic growth in cell numbers. The number of cells dividing is proportional to the number of cells present; the logarithmic phase produces a first order growth in numbers. So the changes in cell numbers correspond to a first order reaction. The numbers cannot increase indefinitely in this way, and they reach a limit (5), which seems to represent more or less a universal 'house full', of somewhere around 10^9 to 10^{10} per ml. Since it is the numbers that most affect quality, it is the logarithmic phase that is critical. Treated as a first order increase, this can be calculated by the reaction rate procedures. For historical reasons, the symbol for the reaction rate constant used for microbiology is μ (the Greek letter mu), but for present purposes it is convenient to use k to emphasise the consistency of processing. The value of k will vary with the species and the environment, and has to be found by experimentation.

The growth has been found to behave generally as shown in Fig. 3.3, displaying at first virtually static numbers for a period known as the lag phase, then a logarithmic growth phase during which cell division occurs at constant time intervals, and finally the growth tails off and numbers decline.

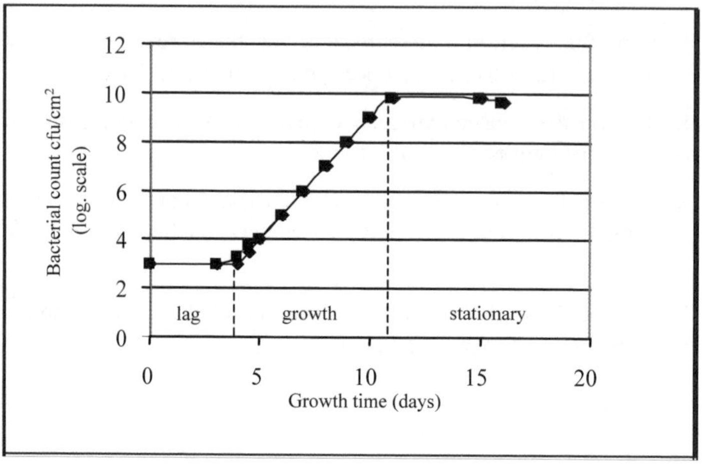

Fig. 3.3. Representative bacterial growth curve

For the logarithmic phase, $\mathbf{kt = ln\ N/N_0}$ (Note the positive sign, with growth)

For the total growth, time = (t + the time for the lag phase). An example of this calculation is shown in Example 3.2.

Example 3.2: Microbial growth in meat wrapped in polythene (from Zamora & Zaritzky (6))

Freshly killed meat is sterile but can be contaminated during the initial processing and storage.

Minimum levels: Some degree of contamination is inevitable but, of course, procedures are rigorously mandated to reduce this to a minimum, which is normally about 10^3 of general colony-forming units (cfu) per square centimetre. At times, it may be slightly lower and the actual incidence depends on both dressing methods and the particular region of the carcass considered.

Maximum levels: The increase in viable organisms depends upon local circumstances, principally temperature, moisture levels, and the gaseous environment, especially if this is deliberately modified, for example by packaging. Spoilage becomes evident with manifestations such as odour, slime, and sometimes discoloration. These start to be apparent at surface microbial counts of about 10^7 - 10^8 cfu cm^{-2}.

In studies of bacterial growth on meat in a polyethylene wrap (giving essentially aerobic conditions because of the high oxygen permeability of thin polyethylene) it was found by Zamora and Zaritzky (6) that *Pseudomonas* species had a constant growth rate of 0.795 day^{-1} and had a lag time of 5 days at 0 °C. Assuming an initial count of $10^{3.5}$ cfu cm^{-2}, calculate the time available before the count will reach 10^8 cfu/cm^{-2} at 0 °C.

Total time to reach 10^8 cfu/cm^{-2} = lag time + logarithmic growth time = t

That is the logarithmic growth "time" = {(total time) – (lag time)}

$$= (t - 5) \text{ days at } 0\ ^\circ\text{C}.$$

$$\ln N/N_O = k(t - 5)$$

and so $\ln(10^8/10^{3.5}) = \ln(10^{4.5}) = 2.303 \times 4.5 = 10.4$

$$= 0.795(t - 5)$$

$$t = (10.4 + 0.795 \times 5) / 0.795 = 14.4/0.795$$

$$= 18 \text{ days}$$

Total available time is 18 days.

There is some evidence that lag times vary with temperature and that the variation can be correlated by an "Arrhenius" type dependency.

Outcomes can be specified at a maximum, shown by deteriorations in the food, such as production of slimes, discoloration or bad odours related to the numbers of microorganisms. This is not usually an exact relationship, and often there is difficulty in setting the upper limit as a microbiological standard. So far as pathogenic organisms are concerned, their incidence is much more arbitrary, as also is any question of an upper end level that can be tolerated. Of course, ideally no level of pathogen is tolerable. But pathogens exist, and they may arise in practical situations. However, there are levels below which they cannot be detected in the food by any present microbiological method. This is about 1 cfu in 25 g of food, corresponding to $4 \times 10^{-2}\ g^{-1}$. In infection and epidemiology studies, there has been work on inoculum levels below which it seems particular organisms cannot establish themselves in hosts (infective levels), but these are only partly definitive, and vary with individuals as well as with their state of health. So working from these is at best difficult, although it is sensible to take them into account when dealing with any defined organism.

A practical way of handling the permitted numbers is to determine experimentally the detection level for the organism, say the $4 \times 10^{-2}\ g^{-1}$, and to set this as a maximum. This gives a starting prescription for handling a difficult and important situation. The question of shelf life can then become one of determining pathogen growth rates, and calculating how long the product would take to reach the minimum measurable contamination level at that rate. This raises the difficult issues of the pathogen source, and the initial numbers involved. Some of these points are discussed in a review paper by Simoni & Labuza (7).

The influence of temperature on microbial growth rates has been studied extensively (8). Growing patterns of different organisms fall into temperature bands, and have been classified according to their temperature ranges for growth: psychrophilic (0-30 °C), mesophilic (10-40 °C) and thermophilic (30-60 °C). The rates of growth increase as temperatures rise within their growth ranges, then decrease and finally cease. A reasonable logarithmic fit, growth rate with temperature, has been found for many organisms over useful ranges of growth, so allowing an Arrhenius type equation; also sensitivities and z values can be used for prediction.

As an alternative for temperature coefficients, a parabolic equation has been suggested labelled the square-root relationship, which corresponds better to some data (8,9) and which has been further elaborated (10). Additional constants in empirical equations can always be introduced to improve fits, but both the precision of the available data and the appropriateness of the applications have also to be taken into account.

For many practical purposes it does not seem critical which is used (11,12). A straightforward constant sensitivity approach provides some feel for the quantities, relates them to other temperature coefficients, and has proved helpful in many practical situations in food processing such as that illustrated in Fig. 3.9. More elaborate treatments can be added as circumstances justify.

3.4.2 Microbial death

When heated and held above the temperature ranges for growth, microorganisms die. Experimentation has shown that the death rates (the number of cells dying per unit time) at any particular temperature are related to, and in fact substantially proportional to, the number of organisms that are viable. Therefore, death rates conform to first order kinetics; at least this is a good approximation over what might be called normal numbers. When numbers become much reduced, then experimental results and any possible rationale become more tenuous and harder both to reach and to justify (13).

Temperature coefficients of death rates have also been found to fit reasonably to logarithmic relationships, so the z and the Arrhenius relationships can be applied. In fact, it was in the context of bacterial death that the z concept was first developed as it seemed a natural measure for the temperature coefficient.

The question of survival is complicated by the need to extend probability ranges right down to considering very small numbers, estimating the chance of retaining one viable organism in large numbers of food entities – for example, cans. This need was shown from the early experimental work. Essentially, the workers started with maximum numbers of viable spores, of the order of 10^{10} to 10^{11} in each of ten tubes. They heated these tubes as quickly as possible, then they measured the minimum time of holding, at a particular constant elevated temperature, after which there was no viable spore found in any of the tubes. This time was termed the 'thermal death time'. Repeating this for the same organism at different temperatures led to a thermal death time curve, relating thermal death time to temperature. The effective reduction in total viable spore numbers was from about 10^{12} to below 1 in the ten tubes, and so the reduction ratio N/N_O was 10^{-12}, and $\log N/N_O = -12$ (or to base *e*, $\ln N/N_O = -12 \times 2.303 = -27.6$).

The concept is analogous to the logarithmic change ratios already seen for other components in a food on processing *following first order reactions*. But there is a decided difference with the magnitudes of bacterial death ratios. For many chemical ingredient components covering a logarithmic (base 10) change ratio of 1 or 2 is sufficient (corresponding to a 90% or a 99% decrease, say, in a component), and 4-5 is larger than most changes encountered in food processing, including microbial growth. In bacterial death, however, the logarithms of such reduction ratios, at least in theory, are 12-15 or even larger. Justification for such wide ranges, and extrapolations, is that, when applied to canning, this early theory not only survived much experimentation, but also seemed to explain observations in canning and provided a useful theory for commercial practice and experience. It has also proved demonstrably safe for most foods.

Current research is active in the areas of continuous processing, in tubular or plate heat exchangers, and where the food contains discrete particles, lumps such as meat balls or mushrooms (14). It is well known that velocity distributions of fluids flowing within a pipe, or between plates, are not uniform. But, even when the fluid is non-Newtonian, a fair guess can be made at the actual velocities and the fastest, where process time is least, quantified (15). The fluid part of the pack can also be temperature/time monitored reasonably easily. What happens inside particles is not so accessible. Even though contamination is generally only on surfaces, it can be inside and there it is difficult for heat to penetrate and even more difficult to be sure that it does so everywhere adequately. Some research investigations have explored this situation, using inoculated food particles (16). Based in part on this work, the US regulatory authorities (USFDA) have quite recently agreed to consider licensing continuous processing under such circumstances, although it is not certain that manufacturers have taken this up. In Europe, limited continuous commercial processing of foods containing particulates (for example mushrooms in a liquid) is in use, and processing specifications seem to be adequate judging from their results.

> *Think break*
> A habit has grown in the food science literature of assessing temperature sensitivities for bacterial deaths in z values and for chemical reactions in activation energies. Discuss the logic and the practicalities of this approach.

3.5 Process Integration

So far we have been considering constant temperature processes, and we have also shown how to put times for real processes, operating at different temperatures, in terms of time for an equivalent extent of processing at a reference temperature. Therefore, we now have the knowledge to deal with normal food processing in which temperatures vary during the course of the process.

3.5.1 General principles

Integration is done by dividing the total process time into sufficiently small pieces during which the temperature is effectively constant, and converting the times for each of these into their equivalent times at the reference temperature. The total time of the whole process is the sum of the individual times. Mathematically, this is equivalent to integration. In fact, integration can be used formally to do exactly the same thing if an analytical expression can be found for the time/temperature relationship (17), but mostly it cannot without rather sweeping approximations being invoked. Although formal integration may seem more elegant, it is no more accurate in practice because of inevitable assumptions that have to be made to enable the mathematical equations to be handled, and it is often no quicker as both can be programmed quite readily using computers.

The basis for the "step" process of integration is shown in Theory 3.2.

Theory 3.2: Basis of the 'step' process for integration

For a 'step' process at a constant temperature:

$$-\int_{Co}^{c} dC/C = \int_{0}^{t} kdt \qquad = kt = k_{T1}\, t_{T1} = k_{T2}\, t_{T2} = \text{constant}$$

and so $t_{T1} = (k_{T2}/k_{T1}).\, t_{T2.}$

This allows calculation of the time at T_1, the reference temperature, equivalent in processing outcome to the actual time at T_2.

Now $k_{T2}/k_{T1} = 10^{(T_2 - T_1)/z}$ by the z method

So for this one step $t_{T1} = t_{T2}10^{(T_2 - T_1)/z}$

If, for example, for a first order process, $f(C) = 1/C$

And choosing $T_1 = 121$ and $z = 10$ (for standard F_0)

we have $t_{121} = t_{T2}10^{(T_2 - 121)/10} = \Delta F_0$ by definition of F_0

where ΔF_0 is the step contribution to F_0

$$\text{And so } (\Delta F_0) = -(1/k_{T1})\, .\int_{Co}^{c} dC/C = -(1/k_{T1}).\ \ln C_1/C_0$$

$$\Sigma(\Delta F_0) = F_0 = -(1/k_{121}).\ [\ln C_1/C_0 + \ln C_2/C_1 + \ln C_3/C_2 + \ldots\ldots]$$

$$= -(1/k_{121}).\ [\ln C_n/C_0]$$

$$= t_{T1}10^{(T_1 - 121)/10} + t_{T2}10^{(T_2 - 121)/10} + t_{T3}10^{(T_3 - 121)/10} + \ldots\ldots$$

taking the process over n steps at temperatures T_1, T_2, ...T_n, which can be chosen to suit particular circumstances.

In words (total F_0) = (sum of "step"F_0 's), each calculated for its individual temperature by the above formula.

3.5.2 Sterilisation/canning

The actual method used for the integration can take many forms. One is to assemble the total F_0 value from piecemeal step contributions, working from calculated times for the reference temperature derived through z value (or Arrhenius) conversions applied to an experimental heat penetration curve. In the canning literature, this has a special name and it is called the General Method. To do it, you need the time/temperature curve, and the kinetic data for the significant reaction(s) in the process – in canning, usually critical spore death.

An example of a simple calculation is shown in Example 3.3.

Example 3.3: An experimental investigation on sterilisation in a can

In the canning experiment shown in Fig. 3.2, the temperatures during the retorting cycle were:

		[Heating cycle	]	[		Cooling cycle			]
Time (min)		3	4	5	6	7	8	9	10	11....
Temperature (°C)										
	Retort	131	134	134	128	121	118	69	44	44....
In can	- upper	118	127	129	131	130	128*	124	104	88....
	- middle	105	117	124	125	125	124	121	91	79....
	- lower	85	105	116	121	121	121	117	74	60....

Calculations were made on the numerical data for each set, working out the equivalent processing time at the standard temperature of 121 °C from the observed time at temperature T with z = 10 by using the relationship

$$k_{121} = k_T\, 10^{(T-121)/10}$$

where k_{121} is the reaction rate constant at 121, which was set at 1, and the T was temperature in degrees C. So for time interval of 1 min at temperature T $\Delta F_0 = k_T/k_{121}$

A sample calculation for the ΔF_0 term, at time 8 min, 128 °C, is shown in Example 3.1.

The relative values of the reaction rate constant at the various temperatures were:

Elapsed time (min)		**3**	**4**	**5**	**6**	**7**	**8**	**9**	**10**	**11**
ΔF_0 (min)										
	Retort	10.0	19.9	19.9	5.0	1.0	0.5	.....	....	....
In can	- upper	0.5	4.0	6.3	10.0	10.0	5.0*	2.0	0.01..	
	- middle	...	0.4	2.0	2.5	2.5	2.0	1.0	....	....
	- lower	...	...	0.3	1.0	1.0	1.0	0.4	....	....

(...indicates regions where data have no appreciable effect on the total integrals)

Contd..

Example 3.3 (contd)

Integration $F_0 = \Sigma(k_T / k_{121}\ \Delta t) = \Sigma\ k_T / k_{121}$

Because Δ t = 1 min, being all 1-min intervals

Therefore $\Delta\ F_0 = k_T / k_{121}$ for each temperature

Adding these component ΔF_O values for each minute of the processing together, over the temperature history, gives the total F_0 of the process for each element of the can.

Therefore F_0 values can: top = **37.8**, middle = **10.4**, lower = **3.7**, and for retort = **56.3**

The numbers have been rounded and the integration is simplified to make the calculation transparent. Both the experimentation and the integration can be improved if more precise figures are required.

The ΔF_O are graphed in Fig. 3.4(a) with line graphs and Fig. 3.4(b) with bar graphs, which also illustrate the integral that is the area below the respective lines.

It is interesting to consider the three measurements of the F_0 criterion, which were taken at the different locations on the centre line in the one representative can during one process cycle: 37.8, 10.4, and 3.7 min. These figures demonstrate clearly the point that, if the process adequacy is determined by the fate of potential spores at the least processed region (in this case with an accomplished measured F_O of 3.7 min.), then other regions, and they could be large proportions of the food volume, can receive a good deal more processing. From the point of view of health hazards they are all safe if the minimum F_O is safe. But other parallel reactions induced by heat at the same time in the can will have continued on average well beyond any optimum. Since these are all almost certainly detrimental to quality, and some may be substantially so, this means that quality suffers.

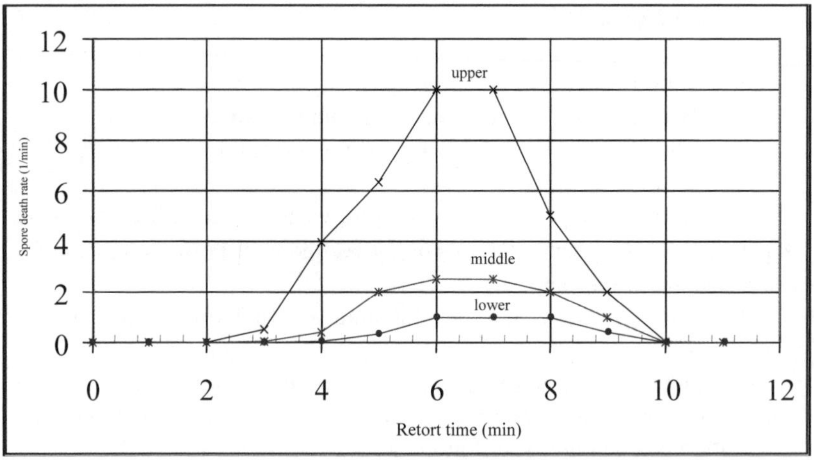

(a) Linear plot

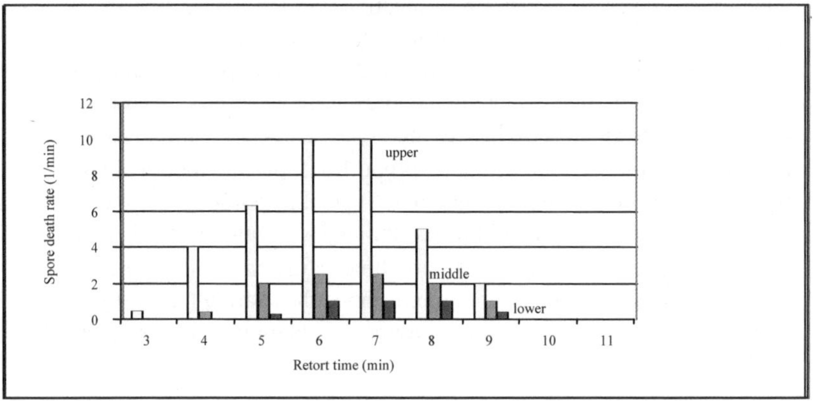

(b) Bar chart

Temperature data from Packer (4)

Fig. 3.4. Rates of spore destruction in an experimental can during processing

One evident problem arises if not all parts of the food being processed have the same time/temperature relationship. Quick reflection will show that, unfortunately, this is true of virtually every practical process, although its impact will vary. Traditionally, this problem has been met by selecting that part of the food that has the lowest F_0 value, the least processed part, and taking the F_0 value for this as also being applicable to the whole food. This assumes that, if all other parts have more processing, they will at least be subject to adequate treatment, and that the critical aspects of the processing are safety and prevention of spoilage, which is true of canning. But it also includes other assumptions that are more equivocal. For example, the theory of canning and sterilisation depends on probabilities, and therefore must include over-probabilities and under-probabilities cancelling each other out. This means that over-processing should be taken into account and offset against under-processing, but the method singles out only the least processed, which may be just a small proportion of the whole. It would not matter if over-processing were positive, or at least neutral to the other product attributes, but almost always it is detrimental to nutrition and to sensory attributes, sometimes seriously so. These problems have been illustrated for canning, but they also apply to other processing.

There are other numerical methods of integration, all essentially equivalent to the so-called General Method. Today, there is computer software to determine F_0 values from a temperature record of the processing. The software includes programming the working on a computer spreadsheet, and is a simple, easy and accurate way to reduce the work of numerical calculations.

In the past, the area under the rate/time graph was measured (counting squares, or using a planimeter, or using the trapezoidal rule or Simpson's method, or cutting out the area and weighing it) to carry out the integration. Unit area on this graph was determined by measuring the area for a known number of units. The appropriate unit area, the value of ∇, which is the integral of $k\{T(t)\}dt$ over that process can then be determined.

A drawback of all of the numerical methods is that they do not incorporate 'feeling' for the sensitivities of the calculated single answer to aspects of the process, which can be controlled by an operator or an instrument controller, and so can be manipulated during the course of the process to give a better product. Sensitivities can be explored through *'what ifs'*, especially accessible on a computer spreadsheet, but it can be rather a hit and miss procedure. Several time/temperature charts or process rate constant/temperature chart can be plotted on the same graph and the extents of processing compared visually.

Another alternative, covered in detail by Ball & Olsen (2), but also in many other publications (e.g. 18) is to fit heating and cooling curves by equations that can be integrated. This operates essentially by converting the temperature data for heating and cooling to a logarithmic form to fit a straight line and then measuring values for the slope (f) and intercepts (J) to define these lines. Situations can be then integrated and particular cases solved by using published tables.

Example 3.3. concerns sterilisation in an experimental static can. Much current large-scale processing involves agitating the cans, for example in hydrostatic cookers, but even in these there will be temperature gradients and therefore variability of processing. This variability leads to overprocessing, which may be substantial, in all regions other than the least processed. To reduce overprcessing, the safety criterion might be altered to work to a defined maximum probability of there being a viable spore in a can, as was discussed many years ago (19). This could also take the level of initial contamination into account rather than just spore reduction ratios, as in the current theory.

Such an approach has never found adequate acceptance. The reason may be, at least in part, that there remain substantial uncertainties in the theory and, therefore, the current factor of safety/ignorance is still considered necessary. Where up to hundreds of millions of cans are produced in each one of many factories annually and sold to large fractions of the population for them to eat, a cautious approach to safety is the only one possible. The canning industry has had an excellent record, using the theory outlined or close variants of it, and working from heat penetration curves applied to the least processed region in the can, however large or small this may be. Such an approach has been demonstrated through massive experience to be safe and robust wherever it has been properly applied, and this is a powerful argument.

3.5.3 Shelf-lives of frozen foods

There is a great deal of data available on the shelf lives of frozen foods, as described in Chapter 1. Results are summarised in many books and conference reports (e.g. 20-22). Comprehensive publications continue to come from the International Institute of Refrigeration (IIR), which include wide-ranging and specialised conference reports with many technical papers, and booklets on 'Recommendations for the Processing and Handling of Frozen Foods' (23), which are updated at intervals.

The straightforward experimental findings for frozen foods suggested an Arrhenius (and also corresponding z type) response, and this is illustrated in Fig. 3.5 (23).

One unique storage problem with frozen foods is their anomalous behaviour at temperatures near to the freezing point. In this region, 0 ° to –5 °C or so, only some of the water present is actually frozen as pure ice, with the remainder being still liquid. Inevitably, with water subtracted in pure ice, the solutes dissolved in the rest become more concentrated. Reactions in this liquid accelerate because reaction rates are higher at the higher concentrations. Lower temperatures systematically decrease the rate constants. The net effect, the resultant of the relative magnitudes of these two opposing influences of temperature and concentration in this critical region can be an increase as the temperature falls. So it is generally most undesirable to dwell in this temperature range any longer than needed for freezing. There may also be special problems with some foods – for example, complex reactions of protein constituents such as enzymes with changing salt concentration and pH, and changes in the size and location of ice crystals, which create mechanical stresses.

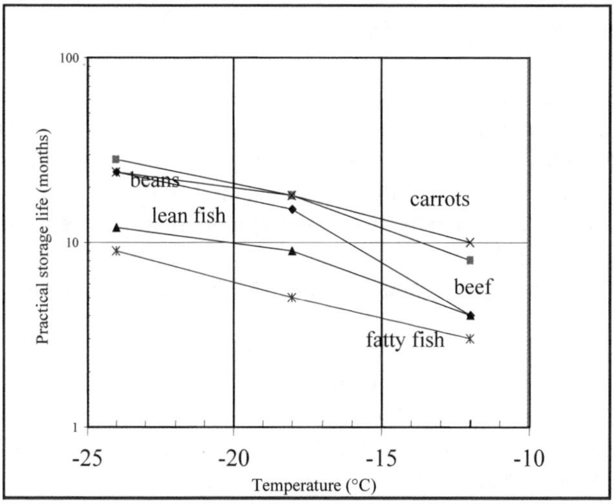

Data recommendations from IIR (21)

Fig. 3.5. Storage lives of representative frozen foods

Otherwise, the general behaviour is sufficiently consistent for the standard reaction technology procedures to be used with reasonable confidence. One that is convenient to use is a variant of the equivalent time concept used for F_0, in the form of fractional changes occurring in times during which temperatures are constant. This assumes *zero order reactions*, a reasonable assumption as the total extent of the reactions is normally relatively small.

For a particular frozen food held at a temperature of T °C, the storage life has been found to be t months; then the average rate of quality loss per month at that temperature is 1/t per month. After x months, the loss is then x/t; after t months the loss is t/t = 1. The fractions of life lost, (x/t), thus determined can be added arithmetically to other corresponding fractions at other temperatures assuming that the storage lives at these temperatures are known, and the resulting sum compared with 1. Less than 1, some storage life still remains; greater than 1 means that the food has deteriorated further than permissible in terms of the acceptable quality measures. This is illustrated in Example 3.4 for the storage life of fatty fish.

Example 3.4: Storage life of fatty fish

A consignment of a fatty fish has been held in storage for 3 months at -20 °C. It is then received into a different store held at -15 °C, and the store operator wishes to have some idea how long it can be held there while still leaving 20% of the storage life for subsequent handlers.

Suggested times for fatty fish to deteriorate to an unacceptable quality are shown in Fig. 3.5.

Reading from Fig. 3.5, and assuming steady deterioration rates, i.e. zero order:

Storage life at -20 °C = 7.5 months.
Fraction of life already expired = 3/7.5 = 40%
Fraction that must be retained = 20%
and so 100 - (40+20) = 40% is still available for the -15 °C store,
Storage life at -15 °C = 6 months

That is (40/100) x 6 = 2.4 months is available for holding in store at -15 °C.

The numbers in the example have been kept simple, but the method has quite extensive application. It is kinetically justified, whatever the order of the deterioration reaction, so long as initial and end 'concentrations', or their corresponding sensory panel assessments, are maintained the same. Only for zero order reactions will intermediate concentrations behave *pro rata* to the percentages, but this seldom matters because there is little interest in intermediate stages and, in any event, discrepancies are normally small. Often it is only the beginning (fresh frozen) and end (total assessed life) points that are important. If the temperature/time line on the OTT chart is straight, the slope of the line (or the corresponding E or z values) can be used for interpolations.

Complex and multi-temperature storage histories can be accommodated so that they can include storage, transportation, and handling. Although the precision of the results is not high, it is as good as that of the actual storage-life data and is adequate for many purposes, such as cold-store stock management, and setting of store temperatures to give enhanced or designed product life where this is practicable.

Storage life experiments are never easy, and, if conducted at storage temperatures, from definition the experimentation takes 'storage' time and in freezer stores this can be years. If the temperature coefficients of the storage deterioration rates can be shown to conform to some defined pattern, Arrhenius for example, then accelerated testing becomes practicable, which can clearly speed things up substantially. Towards the end of the storage life, the measurements become more significant, and Fu & Labuza (11) give an account of a hazard analysis procedure in which sampling is intensified over critical regions by a systematic procedure.

Another plot is shown in Fig. 3.6 taken also from a table in IIR (23), and giving data on high-quality storage lives of frozen fruit in syrup originally from Guadagni (24). All of these show the same general patterns and how shelf lives can be obtained from the literature; the IIR tables cover a great many common products. They also have the advantage of international acceptability, which can be critical at times, especially for export products, and entry into new markets. When data are not available for products of concern, then the same methodology can be used experimentally, but it can be a long task.

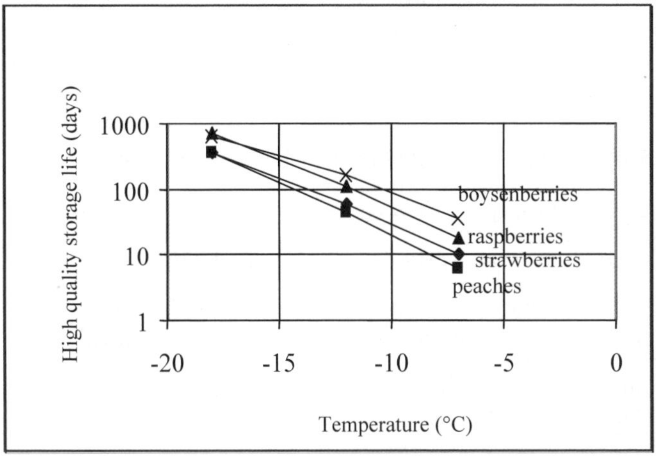

Data from Guadagni (24)

Fig. 3.6. Storage life of frozen fruit in syrup

Most storage reactions are zero order, for ambient, chilled and frozen foods.

Case study 3: Shelf-life of fish

This case study shows how reaction technology can be used to predict shelf lives of foods in a product area, fish. It is worthwhile to explore how seemingly theoretical reaction technology patterns can be applied to predictions that can then be industrially helpful.

Maintaining fish at the highest quality, closest to fresh out of the water after catching, has always been a challenge for food technologists. It has long been recognised that low temperatures at all stages are advantageous, but, because of the expense and practical problems of rapid cooling, the response has continually been 'yes' but how much chilling, and when, and what difference does it make?

Neat tidy categorisation could hardly be expected, because of:

- diversity of species, difference in shape, size and composition,
- fishing all over the world under very varied conditions,
- fish with major variations such as different fat contents, microbial flora and different physiological states such as at spawning,
- different judgement factors that come together to constitute the quality called 'freshness'.

The first problem is how to measure quality? An early measure was the increasing concentrations of trimethylamine, which has a noticeable off-odour and results from bacterial growth and activity in the fish flesh. It was found to rise regularly with time and also to correlate with sensory panel assessments of the condition and deterioration of the fish. Figure 3.7 shows that, over time, the concentration of trimethylamine in the fish increases, but not linearly (data drawn from Lovern, in (25)).

The author commented that for cod and haddock, 'fresh' seemed to imply trimethylamine levels:

- 'Fresh' not more than 1.5 mg/100 g,
- 'Good' not more than 6 mg/100 g,
- 'Poor' not more than 12 mg/100 g; thereafter 'inedible'.

This is not a zero order reaction (or the graph would be a straight line), or a first order reaction. But it smooths to a monotonic increase, which can be fitted reasonably by a second order polynomial, which runs well up into the inedibility range, as shown by the trend line on Fig. 3.7.

Contd..

Case study 3 (contd)

Another measure was consumer acceptability testing, as ultimately it is consumers who determine whether fish for them is fresh or not and have opinions on how fresh (20). Two different studies were compared, about 30 years apart (26 and 27). The later data were composites from a large number (70 sets) of results reported in the literature for protein foods in the range 0 °C to 15 °C, and so seem to have wide applicability. In both studies, freshly caught fish were stored at selected fixed temperatures and for measured times, and then sampled by the sensory panel. Kuprianoff's panel compared stored fish against three quality standards: excellent to very good A, good quality B, and satisfactory C; Bremner's panel judged them for acceptability. The time to reach either the end of the Grade for Kuprianoff or the end of acceptability for Bremner were determined for each storage temperature.

The relative rates of the two sets of data are shown in Table 3.I. The rate at 0 °C was taken as 1 and the other rates were relative to this.

TABLE 3.1
Relative rates of fish deterioration at different temperatures

Temperature (°C)	**0**	**2**	**4**	**6**	**8**	**10**	**12**	**14**
Rate – Kuprianoff (26)	1	1.3	1.6	2.1	2.7	3.4	4.3	5.5
Rate – Bremner *et al.* (27)	1	1.2	1.9	2.5	3.2	4.0	5.0	6.0

Note: rate is the rate relative to the rate at 0 °C, which is taken as 1.

The relationships between the rates at different temperatures are remarkably similar for the two sets of data. These relative rates were plotted against temperature (Fig. 3.8), on both a natural (a) and a logarithmic scale (b) to show how they might be extrapolated.

Consider the implications of the storage life data for fish. Fish on ice (at around 0 °C) have a high-quality life of only about 10 days. It is quite possible on board a catching vessel for fish to be subjected to temperatures between 10 and 20 °C for a substantial number of hours. For example, the relative rate for 10 °C is given as 3.4, that is the fish deteriorates 3.4 times as fast as it would do at 0 °C. Therefore:

- 10 hours at 10 °C deteriorating at 3.4 times the rate at zero (ice temperature) is equivalent to 3.4 x 10 = 34 h on ice.
- If 10 days (240 h) at zero is the high-quality life on ice, then this fish has already lost 34/(240) = 14% of its available high-quality life before it is even stowed in the chiller on board.

Contd..

Case study 3 (contd)

This is a substantial additional quality burden for the distribution system if it is to provide customers with the freshest fish. Alternatively, it places a premium, and a measurable one, on getting that fish temperature down more quickly and holding it down.

The activation energy for fish deterioration was found to be about 78 kJ/mol, and this relates to a sensitivity of about 13% per degree C at around 0 °C. For practical prediction purposes, a reasonably quick estimate of relative deterioration rates at chiller temperatures can therefore be obtained using a constant sensitivity of about 13%/°C over the range.

In fact, the total fish deterioration story is much more complicated and more is known about it now. Deterioration is rapid, it is strongly dependent on temperature and there is only a short high-quality life available. See Fig. 3.9 for an OTT chart for fresh chilled fish, based on experimental results from sensory panels.

The straightforward practical remedy is to maximise life by reducing the temperature. This reaction technology information can be used to assess the value of temperature regimes that might be practicable set against the probable market values of fish either for direct sale or as raw material for subsequent processing. In effect, this analysis reveals two significant quality sensitivities, one chemical and one sensory, and both of these can be used by the fish process technologist to make useful predictions. Other parameters have also been measured: changes in other chemical constituents of the fish, physical changes such as the tensile strength of the fish fibres, biochemical changes in fish proteins, and rates of growth of microorganisms. These may not fit the same reaction orders or temperature sensitivities as noted above. It is not their individual but their overall net impacts that come together to register the consumer panel assessments. From the point of view of industry, individual components may or may not be significant, but the consumers' overall view certainly is.

If the fish is a raw material to be continued into further processing, then it may be one of the constituent properties that becomes critical and so has to be looked at in greater depth. An example is tensile strength, particularly in some species that are inherently particularly sensitive in this regard. Tensile strength is relatively easy to measure and is highly correlated with fish fillet cohesion. In one species, hoki, this tensile strength was found to

Contd..

Case study 3 (contd)

have a somewhat different activation energy, of around 70 kJ/mol, or temperature sensitivity of 12%/°C, making it marginally less sensitive to temperature than the general fish acceptability (28). It indicates greater temperature tolerance. Had it gone the other way with a higher activation energy, however, it could have been critical because the hoki goes mostly to processing, and flesh gaping, a manifestation of tensile weakness, is an important quality issue.

This illustrates that such shifts of quality emphasis arising from different customer needs could indicate different handling requirements. Competitive advantage could hang on appreciating or not appreciating these finely poised situations. So specific measures, as well as the more general acceptability, may justify quite detailed investigations of the processing technology.

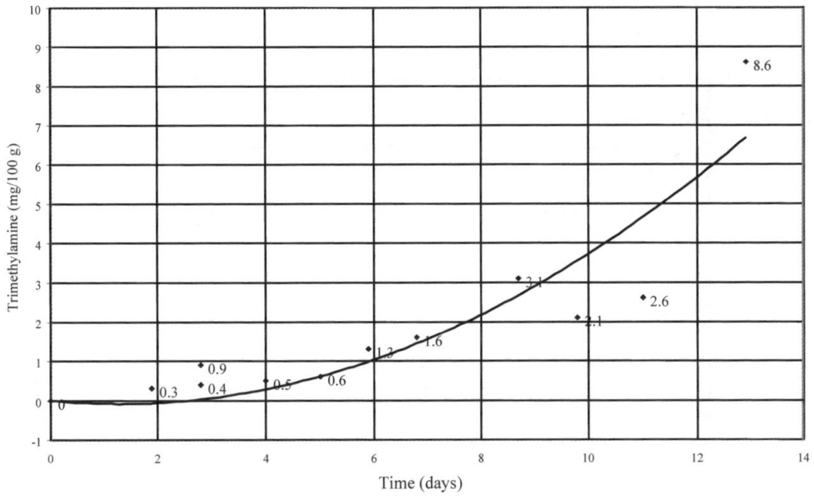

Data from Lovern, quoted in Bate-Smith & Morris (25)

Fig. 3.7. Storage of fish on ice – generation of trimethylamine

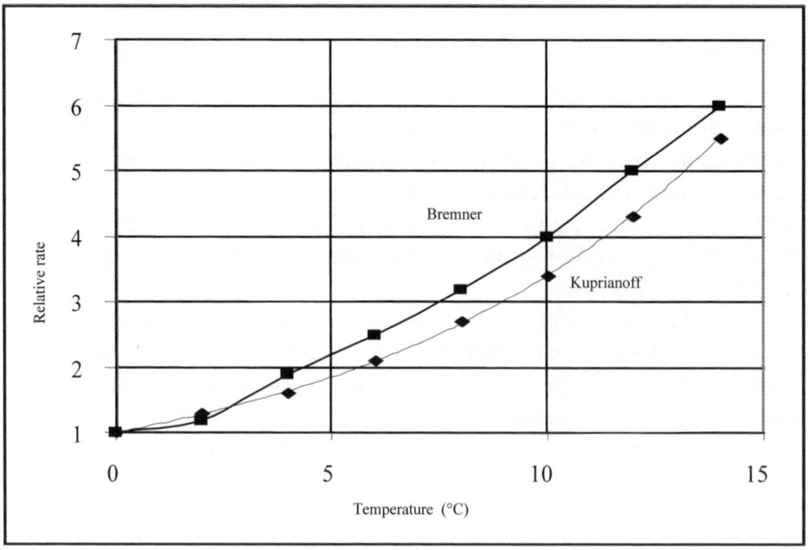

Fig. 3.8(a). Fish deterioration with temperature - natural scale

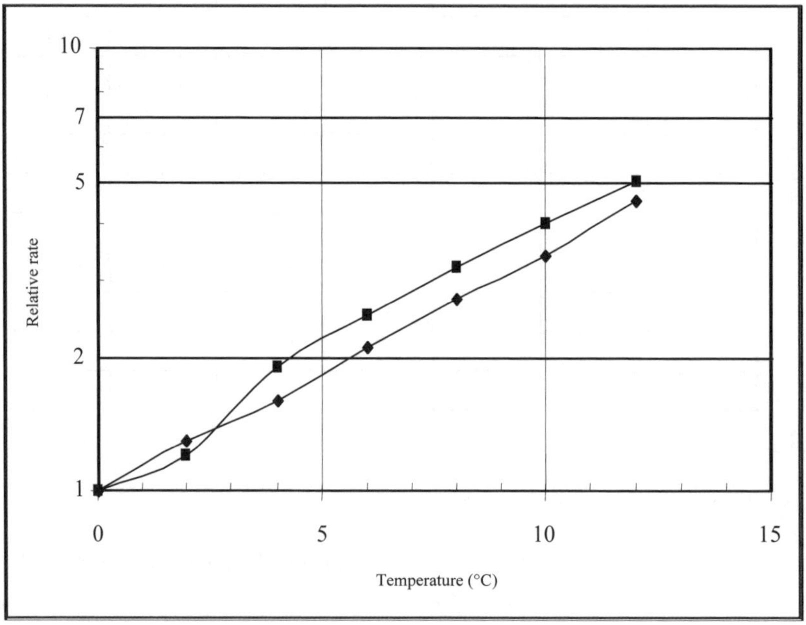

Fig. 3.8(b). Fish deterioration with temperature - logarithmic scale

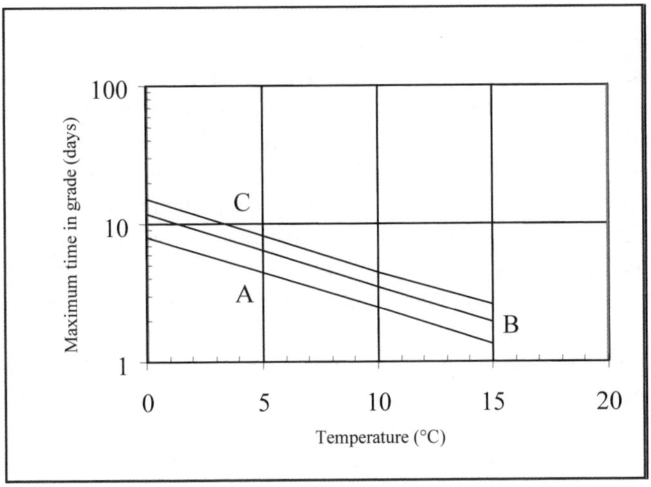

Grades: A = Excellent to very good, B = Good, C = Satisfactory
Data from Kuprianoff (26)

Fig. 3.9. Storage of chilled fish: OTT plot of time and temperature

> *Think break*
> Select appropriate values for the deterioration rates and temperature coefficients for fish after catching:
>
> * Examine the deterioration occurring in the various steps in a typical temperature chain for fish reaching your table.
>
> * Identify the weakest links and suggest practical steps that might be taken to improve quality.

3.6 Practicalities

Applications of the outcome/temperature/time (OTT) charts are numerous, but, in particular, they give quantitative data that can be used to answer directly many important and practical questions with sufficient accuracy for many of the food processors' purposes. They also have wide application, including all manner of ingredients, microbiological aspects, and consumer acceptances. They extend beyond processing conditions studies, to impacts of raw material suitability, and storage lives. Once the data and the charts have been plotted, then application is quite direct and immediate.

3.6.1 *Designing the process*

In designing a process, there are limits to the processing conditions from food regulations, product specifications and plant factors. Within these limits, the process designer can develop an optimum process by using reaction technology. For example, if a set of outcomes such as bacterial reduction ratios of 10^3, 10^5, 10^7,10^9 (conveniently abbreviated to 3D, 5D, 7D, 9D reductions) and intermediate ones, if needed, is plotted on an OTT chart for a particular situation, this can be used to judge processing conditions for particular outcomes. Should an outcome be stipulated, the process conditions can be selected to achieve this.

In heat sterilisation, judgement and experience may suggest that an expected maximum of 1 viable spoilage organism in 10^4 items (packs), after heat processing, provides an adequate shelf life. Average levels are 10^3 per pack in the raw contents on filling. Therefore, any combination of time and temperature on the 7D line ($10^4 + 10^3 = 10^7$, taken down to 1 by 7 decimal reductions) should be adequate. If contamination of the raw ingredients is found to increase to 10^4 per container, the process outcome should move to half way between the 7D and the 9D lines, and the time/temperature combination needs to be altered to achieve this.

Such straightforward calculations do not remove the need for judgement, but they provide a powerful reinforcement tool and an added feeling of confidence when making decisions. The charts can facilitate calculation of F_0 values, and also evaluation of the impact of different temperature coefficients of reaction rates (different activation energies and z values) with all the implications these can have for product quality, and consumer safety and nutrition.

3.6.2 *Controlling the process*

Some questions that can be answered are:

- How precise does the temperature control need to be?
- How precise does the time control need to be?
- How can the process meet the regulatory standards for heating processes?
- How can changing the time and temperature improve the overall standard of the product?

Historically, one important example of this was in the field of pasteurisation of milk. The impetus for heat treatment of milk for drinking came with the knowledge that tuberculosis organisms could be present and that they could be reduced to harmless proportions by heating for a determined time, originally in large batch vats. Then it was found that the same effect on the pathogens could be obtained by heating for shorter times but at higher temperatures and some of the basic quantitative work involved one of the earliest uses of OTT charts in this context to find the correct temperature/time combinations. This led to so-called high-temperature/short-time (HTST) continuous processes. The stipulated requirements were unattainable in large batch vats, so novel equipment had to be invented. What emerged was the plate heat exchanger based originally on the filter press, combining a large heating surface and the possibility of rapid response and good control, with ready access to exposed surfaces for cleaning (1). There were many ramifications, but basically other parallel processes in the milk proceeded less under the higher temperatures and shorter times while still retaining the pathogen reductions required.

Similarly, if the actual working temperature in the plant is a degree or so above normal design, the chart can quickly indicate the impact of this on the colour or the vitamin content or the bacterial reductions, and action taken if it is judged necessary. 'Piecemealing' is also relatively easy, in that part-processing effects can be estimated, and therefore processing additions made that have the impact to produce the necessary quality outcome. So interruptions can be handled with greater assurance. A widely extending use is when, as is almost always the case, the many reactions that are proceeding simultaneously have important consequences for the quality of the product and so the best obtainable trade-offs have to be sought and secured.

3.6.3 Benefits of outcome/time-temperature charts

Combined effects of time and temperature can be explored systematically on the relevant OTT charts.

Experience, over many years, exemplified by milk pasteurisation investigations, has shown that OTT charts have provided:

- New and deeper insights into an important and very widely used process.
- Better quality product with easier and more certain bacterial reduction.
- Methodology that increased understanding of process systems with wide application.
- Possibility of extension from fixed temperature to variable temperature and continuous operations, and therefore much wider process design scope.
- Simplicity of understanding, and ready manipulation, of data and conditions.
- Insights and drive for new equipment leading to process hardware innovation.
- Gradual recognition in regulations that overall process outcome, as opposed to step details, could be equally effective and less restrictive, yet leave much more scope for process development.

3.6.4 Relating outcomes to process conditions

Another important insight given by the charts is the effect on the outcome (the product) of different degrees of treatment because of differentiated time/temperature conditions arising inevitably within different parts of the product. This has been briefly examined with respect to contents located in different parts of a can, but its impact is very widespread indeed in foods being processed. It can sometimes be possible to alter the conditions, or the packing, or the flow patterns to improve uniformity. Of course, sometimes non-uniformity is desirable. But, in any event, the OTT charts provide a useful tool to explore, quantitatively if the relevant different temperature histories are known, just what are the consequences of non-uniformity in the processing in the various regions of the food materials.

Overall, the great advantage of the OTT charts is the amount of process information they can contain compactly, combined with accessibility and adaptability. So that it could be very worthwhile to investigate their availability in particular situations and to undertake the modest amount of effort involved in preparing such charts and gaining familiarity with them.

> *Think break*
> Using appropriate values give your answers to the questions posed in 3.6.2 for a canning process and a milk pasteurisation process.
> To what extent are these questions covered or accommodated in regulatory regimes?

3.7 References

1. Lewis M.J., Heppell N.J. *Continuous Thermal Processing Of Foods.* Gaithersburg. Aspen, 2000.
2. Ball C.O., Olsen F.C.W. *Sterilization in Food Technology.* New York. McGraw-Hill, 1957.
3. International Critical Tables. New York. McGraw-Hill, 1927.
4. Packer G.J.K. *The Development of a Chemical Analogue for Thermal Destruction of Bacterial Spores.* Thesis, Massey University, Palmerson North, 1967.
5. Labuza T.P., Fu B. Growth kinetics for shelf-life prediction: theory and practice. *Journal of Industrial Microbiology, 1993,* 12, 309.
6. Zamora M.C., Zaritzky N.E. Modelling of microbial growth in refrigerated packaged beef. *Journal of Food Science,* 1985, 50, 1003.
7. Shimoni E., Labuza T.P. Modelling pathogen growth in meat products: future challenges. *Trends in Food Science and Technology,* 2000, 11, 394.

8. Zwietering M.H., de Koos J.T., Hasenack B.E., de Wit J.C., van't Riet K. Modelling of bacterial growth as a function of temperature. *Applied and Environmental Microbiology,* 1991, 57, 1994.

9. Ratkowsky D.A., Olley J., McMeekin T.A., Ball A. Relationship between temperature and growth rate of bacterial cultures. *Journal of Bacteriology,* 1982, 149, 1.

10. Rosso L, Bajard S., Flandrois J.P., Lahellec C., Fournaud J., Veit P. Differential growth of *Listeria Monocytogenes* at 4 C and 8 C: consequences for the shelf life of chilled products. *Journal of Food Protection,* 1996, 59, 944.

11. Fu B., Labuza T.P. Shelf-life testing: procedures and prediction methods, in *Quality in Frozen Food,* edited by Erickson M.C., Hung Y-C. New York. Chapman & Hall, 1997.

12. Buchanan R.L., Whiting R.C., Damert W.C. When is simple good enough: a comparison of the Gompertz, Baranyi, and three-phase linear models for fitting bacterial growth curves. *Food Microbiology,* 1997, 14, 313.

13. Peleg M., Cole M.B. Reinterpretation of microbial survivor curves. *Critical Reviews in Food Science and Nutrition*, 1997, 38, 353.

14. Ramaswamy H.S., Awuah G.B., Simpson B.K. Heat transfer and lethality considerations in aseptic processing of liquid/particle mixtures: a review. *Critical Reviews of Food Science and Nutrition,* 1997, 37, 253.

15. Simpson S.G., Williams M.C. An analysis of high temperature/short time sterilization during laminar flow. *Journal of Food Science*, 1974, 39, 1047.

16. Palaniappan S., Sizer C.E. Aseptic process validated. *Food Technology*, 1997, 51 (8), 60.

17. Deindoerfer F.H., Humphrey A.E. Analytical method for calculating heat sterilization times. *Applied Microbiology,*1959, 5, 256.

18. Toledo R.T. *Fundamentals of Food Process Engineering.* 2nd Edition New York. Van Nostrand Reinhold, 1991.

19. Hicks E.W. On the evaluation of the canning process. *Food Technology,* 1951, 5, 134.

20. Jul M. *The Quality of Frozen Foods*. London. Academic, 1984.

21. Erickson M.C., Hung Y-C. (Eds). *Quality in Frozen Food.* New York. Chapman & Hall, 1997.

22. Kennedy C.J. (Ed.) *Managing Frozen Foods.* Cambridge. Woodhead, 2000.

23. International Institute of Refrigeration (IIR). *Recommendations for the Processing and Handling of Frozen Foods.* 3rd Edn. Paris – International Institute of Refrigeration, 1986.

24. Guadgagni D.G. Quality and stability in frozen fruits and juices, in *Quality and Stability of Frozen Foods*, edited by Van Arsdel W.B., Copely M.J., Olson R.L. New York. Wiley Interscience, 1969.

25. Bate-Smith E.C., Morris T.M. Food Science: A Symposium on Quality and Preservation of Foods. Cambridge. Cambridge Press, 1952.

26. Kuprianoff J. The effect of temperature and the duration of storage on the changes in foods during frozen storage. *Kaltetechnik,* 1956, 8, 103.

27. Bremner H.A., Olley, J., Vail A.M.A. Estimating time-temperature effects by a rapid systematic sensory method, in *Seafood Quality Determination*, edited by Kramer D.E., Liston J. Amsterdam. Elsevier, 1986.

28. MacDonald G.A., Stevens J., Lanier, T.C. Characterization of hoki and Southern Blue whiting compared to Alaska pollock surimi. *Journal of Aquatic and Food Production Technology*, 1994, 3, 31.

4. ACHIEVING BETTER FOOD PRODUCTS

4.1 Introduction

To apply reaction technology in practical processing and storage, a broad range of factors may need to be considered. So far, only one basic model of rate prediction has been considered, and it has been shown that this fits many commonly encountered and practical food processes. But, for some important food processing systems, the simple power law relationship that was introduced in Chapter 2 does not fit experimental results and practical situations sufficiently. Some further elaboration of the model is required to cope with the needs of processing and process control. Fortunately, fairly straightforward extensions can be used to explore many of these situations, to either adapt or confirm the model. Two important aspects are firstly the changes in reaction rates during a process, and secondly the presence of multiple reactions in many food processes.

Some reaction patterns and apparent orders of reactions change as the reactions proceed. Changing process conditions, particularly temperature, can cause these. There can be complex changes in the nature of enzyme-catalysed reactions, because of denaturation of the proteins. These changes can be built into the whole kinetic analysis. It may prove adequate to consider only part of the whole reaction spectrum because that part is enough to cover the situations in practice.

Multiple reactions are often important in food processing – both chains in which a series of reactions follow one on the other, and also parallel reactions in which a number of reactions occurs at the same time. Reactions can be alternative paths from the same initial reactants, leading to different products that may have very different levels of desirability in the food. Or they may be reactions that are independent, but linked by identical process conditions that accelerate all the reactions. Because of the multiple constituents of the food, processing conditions, such as a temperature rise, will simultaneously be applied to all the components present. Therefore, these components will change and some of their changes may not be desirable. So conditions have to be sought that accomplish the required process change on one critical component or attribute because it is the primary or dominant reason for the processing but minimise or modify the simultaneous changes in important components or attributes. For example, safety may be a critical product attribute, but the texture may be important. The process conditions need to satisfy the critical attribute, but also optimise the important attributes.

Such possibilities can be found from consideration of the changes and the kinetics. Optimum ways of controlling reactions and obtaining the best overall product can then be explored through analysis by the reaction technology.

4.2 Changing Reaction Rates

Changes in reaction rate constants can be caused by changes in the processing conditions – for example temperature. Changes in reaction rate constants can also occur in enzyme-catalysed reactions. Shelf storage changes often follow apparently zero order, which changes to first order with time. Concentration levels thus seem to influence not only the reaction rate constant but also the observed order of some processing reactions.

4.2.1 Changes in temperature

In the rate equations considered so far, it is implied that they cover the whole concentration range quite irrespective of whether all, or only part, of that range is encountered by the food processor. However, the food processor is interested only in the range needed for the processing; what goes on outside that may be interesting theoretically but is not of importance to practice. This can give scope for considerable simplification; for example over limited concentration ranges it may be possible to use zero order, although overall the reaction may be of a higher order. In shelf-life calculations and the product-life dating that stems from them, only zero order, that is constant reaction rate, needs generally to be taken into account and this still retains accuracy adequate for day-to-day store and warehouse practice.

Storage often involves changing temperatures, which means changing reaction rate constants. An example of changing storage conditions at common ambient conditions in temperate countries (15-25 °C) is shown in Example 4.1, which is using the reaction technology data in Case Study 2 to predict the storage life of the whey-coated confectionery products. The yellowing was found to follow zero order by Trezza & Krotcha (1).

Example 4.1: Whey protein coated confectionery: calculation of shelf-life over different holding temperatures

In Case Study 2, the yellowing of whey protein concentrate coated confectionery was zero order, and it is illustrated in Fig. 2.8.

If, after storage for 6 months at 15 °C, the confectionery were transferred to a retail store at working temperature of 25 °C, for what maximum time should it remain on the shelf if yellowing is not to exceed 15 Yellowness index units?

This can be calculated by determining the monthly rate of yellowing at the two temperatures of 15 °C and 25 °C, and then the fractional loss of total storage life in each.

The monthly reaction rate constant (k) for the yellowing of the coating can be taken at 23 °C as 0.29 index units/month, with corresponding activation energy as 95 kJ/mol.

Rate of deterioration at 23 °C, k_{23}, is 0.29 index units/month,

(15 °C=288 K, 23 °C=296 K, 25 °C=298 K)

$k_2 = k_1 \exp\{-95000 \text{ x } (T_1 - T_2) /(RT_1 \, T_2)\}$ where T_1 = 23 °C, T_2 = 15 °C

so at 15 °C, k_{15} $= 0.29 \exp\{-(95000 \text{ x } 8)/(8.314 \text{ x } 296 \text{ x } 288)\}$
$= 0.29 \text{ x } 0.342 = 0.1$ index units/month

So, after 6 months, 0.1 x 6 = 60% of its available life would have been lost, 40% left.

at 25 °C, k_{25} $= 0.29 \exp\{-(95000 \text{ x } -2)/(8.314 \text{ x } 296 \text{ x } 298)\}$
$= 0.29 \text{ x } 1.3 = 0.38$ index units/month

And so in 1 more month the remaining storage life fraction of 40% would essentially be gone at a loss rate of 0.38/month, i.e. 38%/month.

If there were to be other further calculations on the same material, it would be routinely easier and adequate for most purposes if the sensitivity were calculated at the mean temperature of about 20 °C and used as below:

Sensitivity = $\exp\{95000/(8.314 \text{ x } 293 \text{ x } 293)\}-1 = 0.14$

for example, reaction rate constant at 15 °C = $0.29\,(1.14)^{-8} = 0.1$ index units/month.

In Example 4.2, the decrease in concentration of the sweetener aspartame, during storage was studied in the ambient range of temperature under tropical conditions from 20 °C to 55 °C. This was found to be a first order reaction. The example shows that reactions determining shelf life are not always zero order, and there is a need to check in order to study them for order.

Example 4.2: Shelf-life of a sweetener, aspartame

One of the sweeteners used in soft drink manufacture is aspartame, which is built up from phenylalanine and aspartic acid. Aspartame has a relatively short shelf life in solution, and obviously this is of importance to the extensive industrial use of this sweetener.

The rates of this degradation in a solution with caramel colouring were investigated by Wang & Schroeder (2). They reported a first order degradation reaction over the range of their experiments, 20 °C to 55 °C. The reaction rate constants were:

55 °C	0.0786 d^{-1}	45 °C	0.0277 d^{-1}
35 °C	0.0119 d^{-1}	20 °C	0.00267 d^{-1}

If a quantity of this solution is stored for 3 weeks at 40 °C and then for a further 5 weeks at 25 °C, what fraction of the aspartame originally made up into the solution would be expected still to be present?

An Arrhenius plot of the rate data, plotting ln (reaction rate constant) against 1/T is shown in Fig. 4.1.

The slope from this gives an activation energy of 77 kJ/mol, so that working from the 35 °C (308 K) rate of 0.0119 d^{-1} gives:

At 40 °C (313 K) k_{40} = 0.0119 exp{(-77000 x -5)/(8.314 x 308 x 313)} = 0.0192 d^{-1}

At 25 °C (298 K) k_{25} = 0.0119 exp {(-77000 x 10)/(8.314 x 308 x 298)}= 0.00433 d^{-1}

$t = (-1/k) \ln(C/C_0)$

And so on storage for 3 weeks at 40 °C:

$3 \times 7 = (-1/0.0192) \times \ln(C/C_0)$

$C/C_0 = 0.67$ so about one third (33%) has been lost.

And on further storage for 5 weeks at 25 °C:

$5 \times 7 = (-1/0.00433) \times \ln(C/C_0)$

$C/C_0 = 0.86$ so 14% has been lost.

In total, after the two periods: the residual content would be 0.67 x 0.86 = 0.58 or about 42% lost in all.

The experiment was repeated with different levels of caramel up to 700 ppm, and it was found that reaction rate constants were approximately the same, but above this level of caramel the rate of reaction increased.

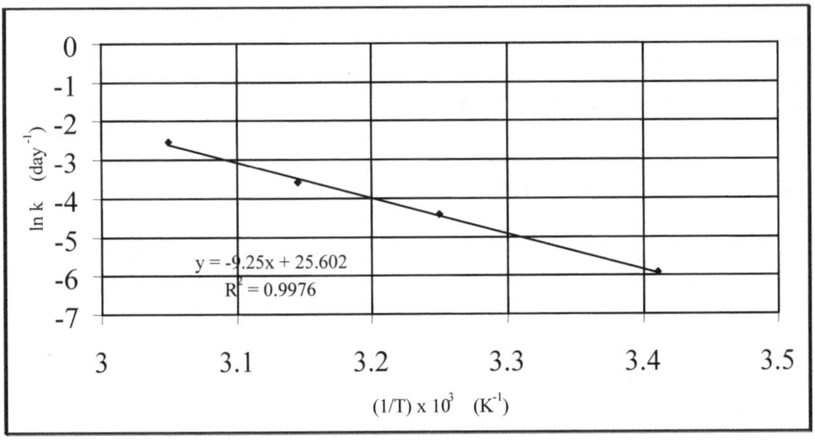

Data from Wang & Schroeder (2)

Fig. 4.1. Loss of aspartame on storage: Arrhenius plot

> *Think break*
> Consider two reactions that can be important in storage tests on a tomato salad dressing: fat oxidation and loss of red pigments (e.g. lycopene):
>
> * How does altering the temperature of storage affect these reactions?
> * How might these reaction changes significantly alter the resulting product composition and product attributes?
> * What other storage conditions could be altered to affect the rates of the reactions?
>
> *Note: in the Think breaks in this chapter, you may have to find the information from textbooks and data bases.*

4.2.2 Enzymic-catalysed reactions

Some very important reactions occurring in food processing are initiated by the presence, and modified by the concentration of non-reacting catalysts. These cause the reaction to occur but do not themselves undergo change as the reaction proceeds. The catalysts include chemical constituents, one being hydrogen ions whose concentration is measured by pH. They include enzymes that are biological catalyst proteins, which can affect the rate of reaction, but may themselves be denatured by the processing conditions. Sometimes the enzymes can be used for promoting desired changes, for example in coagulation of milk or tenderising of meat or softening the centres of chocolates. But other times they may promote undesired deterioration, in which case processing may be employed to substantially eliminate this. So inactivation of the enzymes by heat denaturation may be the important reaction, for example in the blanching of vegetables. The blanching of carrots is shown in Example 4.3.

Example 4.3: Enzymic action in carrots

Blanching to inactivate enzymes is common in vegetable processing. In a study of blanching of carrots, Roy *et al.* (3) reported that the removal of the enzyme, lipoxygenase, was a first order reaction. Blanching treatments in water for 71 min at 70 °C, 11.64 min at 80 °C, 2.12 min at 90 °C and 0.58 min at 100 °C resulted in the reduction of enzyme activity by 80%.

These results have been plotted in Fig. 4.2.

From this Arrhenius time plot, the activation energy for the enzyme destruction was estimated, as 167.5 kJ/mol.

The reduction in the activity of the enzyme, lipoxygenase, was specified as 80%, from which the reaction rate constant could be worked out. The resulting calculation of the activation energy, 167.5 kJ/mol, was true for the range of concentration from 100% to 20%.

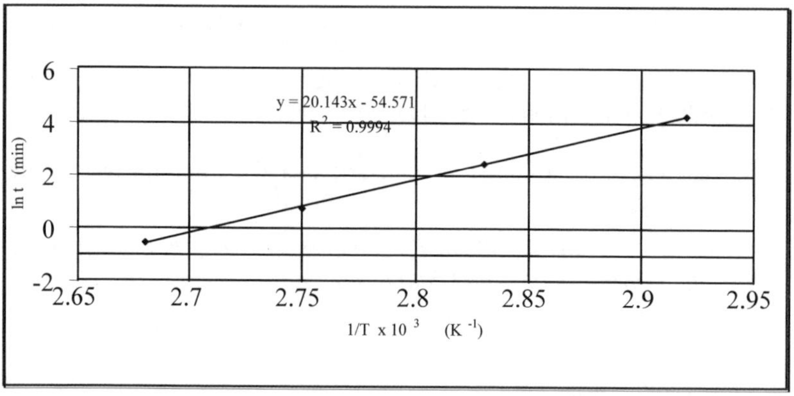

Data from Roy *et al.* (3)

Fig. 4.2. Enzymic reduction in hot-water blanching of carrots

It has been found that many enzyme-catalysed reactions proceed at rates that can be characterised by an equation of the form

$$-r_A = k_1 C_A / (1 + k_2 C_A)$$

where k_1 is the reaction rate constant for the enzyme-substrate reaction (the substrate being the food constituent that is reacting), k_2 *is* the reaction rate constant of (heat) denaturation of the enzyme and C_A is the concentration of the substrate. The equation arises from the net reaction rate constant being related directly to the enzyme/substrate reaction rate constant, and inversely to the enzyme activity-loss reaction rate constant.

On examination of this equation, its apparent order is determined by the relative magnitudes of the two terms in the denominator. When 1 is large compared with k_2C_A, then the denominator is effectively 1 and the overall rate is approximately first order, whereas when k_2C_A is large compared with 1, then the apparent order is zero as the 1 becomes negligible and the C_A cancels out from above and below the line. This implies that, at large concentrations C_A of the substrate, the effective rate is zero order. But when the substrate gets down towards disappearing and the products dominate, then the net order becomes first order. Rather than the substrate concentration becoming zero, it instead diminishes exponentially.

Another important outcome of such kinetics, more directly related to enzymes, arises from the relative magnitudes of the activation energies found for enzyme/substrate reactions and enzyme denaturation reactions; usually the activation energy for the enzyme denaturation reaction is much greater than that for the enzyme/substrate reactions. For example, the enzyme/substrate reaction can be as low as 20, but it is usually around 100 kJ/mol; and enzyme denaturation reactions are usually upwards of 150 kJ/mol. This focuses on the k_2 rather than C_A, but again the effect is to shift the product $k_2\ C_A$ relative to 1. Because of the high activation energy incorporated in k_2, this shifts the overall rate quite dramatically with changing temperatures.

This explanation accounts for the well-known effects of temperature on an enzymic food processing operation, a common example being encountered in the formation of curds from milk when clotted with the enzyme rennin. Here, as the temperature rises through 30 and 40 °C, the clotting (milk protein denaturation) proceeds ever more rapidly, which is the reason why the vats are heated in cheese making. However, if the temperature continues to rise above 50 °C, the clotting rate decreases, slowly at first and then much more rapidly as temperatures continue to increase because of the denaturation of the enzyme.

Think break

Two food processes in which enzymes are significant are the manufacture of soft-centred chocolates and the production of maltodextrins and glucose syrups from starch.

* What are the enzymes and the related reactions for each of these processes?
* In what ways would the process reactions be altered if the processing conditions, particularly temperature, were modified?
* How would these changes affect the final food products?

One enzymic reaction that has been explored in detail is the decomposition of hydrogen peroxide by peroxidase enzymes, such as catalase that can be obtained from liver. To gauge the rate of the reaction, it is convenient to measure the evolution of the oxygen, which can be done by simple collection in a volumetric cylinder. Early results from such an experiment are plotted in Fig. 4.3, where the logarithm of the reaction rate is plotted against the reciprocal of the absolute temperature in the usual way.

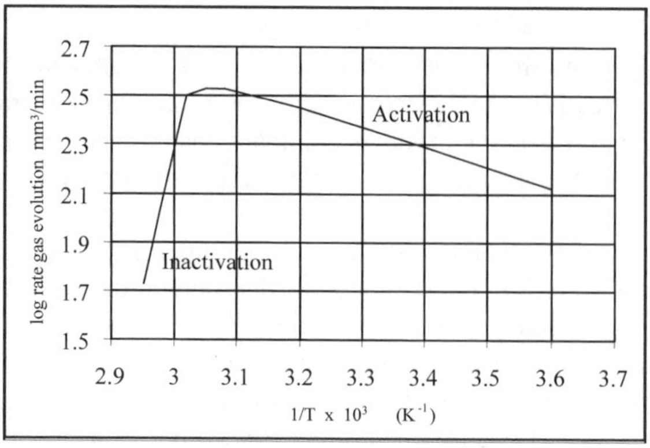

Data from Aiba *et al.* (4)

Fig. 4.3. Decomposition of hydrogen peroxide by catalase: Arrhenius plot

Looking at Fig. 4.3, and recalling that, because reciprocals are being plotted, temperature is highest on the left and falls to the right, it can be clearly seen to fit into two sections. The one that is on the left has a very steep slope and therefore very high value of E, which in this case can be called "inactivation" energy. On the right is a more gentle slope corresponding to a much lower activation energy, which corresponds to the decomposition reaction of the hydrogen peroxide speeding up in the usual way as the temperature increases.

This is the typical pattern, and can be used in processing by heating to initiate the enzymic action, continuing heating to speed it up, but then terminating the action by a still further rise in the temperature and so inactivating the enzyme and ending activity. Between the two defined regions lies a "turnover" temperature range, which is not very well defined but in which the reaction rate is at a maximum. Within the range of possibilities, temperatures can be manipulated to suit processing requirements. An equation of the enzymic form can be fitted, both in practice and in theory. For hydrogen peroxide decomposition, in this equation, k_1 in the numerator corresponds to the activation energy of the enzyme/substrate reaction (17.6 kJ/mol), and k_2 in the denominator corresponds to the inactivation energy of the enzyme denaturation (230 kJ/mol) (4).

Kinetic understanding of this behaviour can also be used to prescribe heating rates that need to be applied in processes such as blanching. This is primarily to destroy enzymes that react deleteriously with the foodstuff, but, if the heating is too slow, there will be time for the enzyme to cause substantial deterioration before being sufficiently denatured.

4.3 Sequential (Chain) Reactions

In some reactions encountered in food processing, the first products of the reaction are themselves unstable, and, as the reaction-promoting conditions of the processing continue, they, initially products, become themselves reactants and form further products. This process can continue, and so in the final food product a whole string of components from the original reactant can arise. Some of these may be desirable, some less desirable, and some undesirable, and both their presence and their relative concentrations can be of considerable significance in the final food product. The craftsman recognises these changes and adjusts processing to optimise the product that is wanted. The technologist seeks understanding both to obtain the specific levels of the desired attributes in the final product, and also to use instrumental signals to control process conditions to adapt to changing inputs so that the specific product attributes are achieved and maintained.

As might be anticipated from the diversity of reactions, each with its own reaction rate constants and activation energies, and each dependent on upstream and downstream situations, the analysis of chains of reactions becomes complicated. However, consideration of the simplest situation with one initial reactant, A, moving with rate constant k_1 to one intermediate product, B, moving at rate constant k_2 to one final product, C, demonstrates the patterns to be expected and provides a useful guide as to what process manipulation might be possible. Assuming that, at the outset of the reaction, at time 0, neither of the products has been formed, and assuming all reactions are first order, after time t the concentrations of the three components are given by:

$$C_A = C_{AO}\, exp(-k_1\, t)$$
$$C_B = C_{AO}\, k_1\{ exp(-k_1\, t) - exp(-k_2\, t)\}/ (k_2 - k_1)$$
$$C_C = C_{AO} - C_A - C_B$$

These equations for sequential reactions give rise to graphs of the form shown in Fig. 4.4.

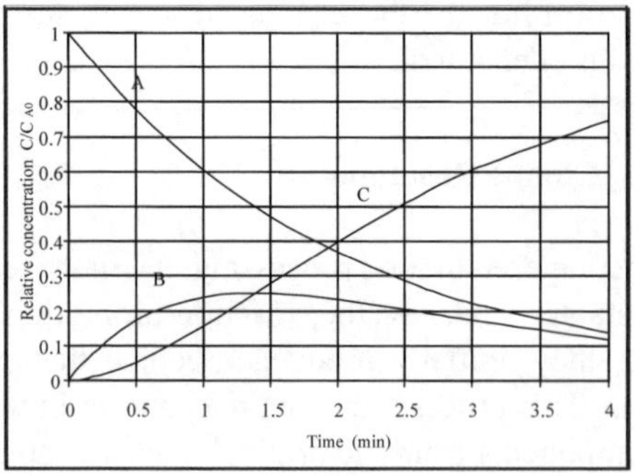

Fig. 4.4. Behaviour of components in a chain reaction

This shows that, as time progresses, the concentration of the initial reactant, A, decays exponentially, whilst the concentration of the intermediate, B, starts from 0 and then rises to a maximum before declining finally to 0. The final product, C, starts at 0 and thereafter rises smoothly, ultimately reaching asymptotically a value equal to C_{AO}, by which time both A and B have disappeared. An outline of the derivation of the equations is given in Theory 4.1.

Theory 4.1: Consecutive (chain) reactions

The equations for the simplest set of consecutive reactions A→B→C, with first order rate constants k_1 and k_2 respectively, can be written:

$$r_A = d\,C_A / dt = -\,k_1 C_A$$

$$r_B = dC_B / dt = k_1 C_A - k_2 C_B \text{ and}$$

$$r_C = k_2 C_B$$

The first equation is familiar and leads to

$C_A = C_{A0}\, exp(-\,k_1 t)$ where C_{A0} is the initial value of C_A

This value can then be substituted into the second equation, which leads to:

$$dC_B/dt + k_2 C_B = k_1 C_A = k_1 C_{A0}\, exp(-k_1 t)$$

This equation can be solved for C_B by the mathematical technique of multiplying by an integrating factor to make the left-hand side into the differential of a product, which can then be integrated. Without going into the detail, which can be found in textbooks on differential equations, this leads to the solution, assuming that at time 0 there is neither *B* nor *C*:

$$C_B = k_1 C_{A0}\, [\, exp(-k_1 t) - exp(-k_2 t)\,] / (k_2 - k_1)$$

and $C_C = C_{A0} - C_A - C_B$ where we now know C_{A0}, C_A, and C_B

The equations indicate how the balances of the components change with time and with the particular rate constants. Consideration of the effect of the activation energies of the k's in the sequential reactions shows that the intermediate component concentrations, maxima and relative, can be changed by altering temperatures. This gives the processor a powerful tool, manipulation of times and temperatures to reach optimum products.

The effect of changing the ratio k_1 / k_2 can be demonstrated by inserting typical values into the various k's and activation energies. The shape of the curves change as the ratio of the reaction rate constants change. If k_1 / k_2 is very large, then effectively A becomes B almost instantaneously, the processing time is occupied by the change of B into C, and the process shape and rate are dominated by the value of k_2. Conversely, if k_1 / k_2 is very small, the processing is dominated by the transformation of A, and there is little or no B to be found at any time in the mixture, as immediately B is formed it reacts to become C. Essentially, A reacts at a rate determined by k_1 to form C. In between these extreme ratios, especially if B is a wanted product, process times and ratios of the k's can be selected to suit to a limited but still significant extent.

A number of examples of sequential reactions are found in food processing and storage. In the Maillard browning reaction (non-enzymic), the intermediate products are many and diverse and they can polymerise, changing from colourless to yellow and finally to a sticky brown mixture. Important to the processor in the case of the browning reaction are the flavours, colours and nutritional significance, of the intermediate compounds, and the relative extents to which these are formed. An intermediate product that can be used to monitor browning is hydroxymethyl furfural (HMF), but often colour measurement is used to follow browning development. That the reaction equations can be built up into broad but quite detailed kinetic schemes that can handle and simplify even the very complex systems such as the Maillard reaction system is described by Jousse *et al.* (5). They classified the volatile compounds, which are the basis of browning flavour. In a kinetic scheme with 11 reaction steps, they were able to correlate pseudo-first order rates of generation of these compounds, which predicted the actual build-up of flavour through processing with the temperature and concentration of the reactants. The equations were then integrated to follow the build-up of flavour. Although the mathematics involved 11 differential equations that had to be solved simultaneously, once the equations had been written and parameters such as rate constants, concentrations and temperatures inserted, the numerical solutions were produced through standard computer programs.

> *Think break*
> Select a chain reaction system occurring in food processing – for example, a browning sequence in the manufacture of toffee or of caramel:
>
> * Consider how important consumer characteristics of the food could be altered by modifying the timing sequence and, or, the temperatures of the operating sequences.
>
> * To what extent might knowledge of the kinetic sequences improve operation or control of the processing?

Unsaturated fats found naturally are generally softer than corresponding fats with the same carbon chain length but higher degree of saturation. If such fats are to be used in spreads that require firmness, it is desirable to increase the degree of saturation. This can be accomplished by catalytic hydrogenation. In the hydrogenation (catalytic) of unsaturated fats, the intermediate components are fats of progressively lower unsaturation, and the final fats would be completely saturated if the reaction continued to this extent. In the case of fat hydrogenation, important measures of the reaction are:

- degree of residual unsaturation
- hardness of the fat, which rises as saturation rises and which is the reason for conducting the reaction industrially
- nutritional value, where higher unsaturation is generally preferable.

An example of multiple double bonds being progressively saturated is found in the sequence of 18-carbon fatty acids, with linolenic acid (3 double bonds), to linoleic acid (2 double bonds) to oleic acid (1 double bond), to stearic acid, completely saturated. In fats this is related to progressive hardening. Because of their commercial importance, these reactions have been extensively studied. One study is described in Example 4.4.

Example 4.4: Hydrogenation of soya bean oil

In one report, Chen *et al.* (6) reported values for the first stages in hydrogenation, that is linolenic to linoleic, and linoleic to oleic. These were both first order reactions.

The reaction rate constants at 200 °C under one particular set of experimental conditions were:

Linolenic to linoleic	$k_1 = 0.25\ \text{min}^{-1}$ and an activation energy of 44kJ/mol
Linoleic to oleic	$k_2 = 0.42\ \text{min}^{-1}$ and an activation energy of 48.5 kJ/mol

Consider the first three acids in the sequence, linolenic A, linoleic B, and oleic C. These lead through the equations for sequential reactions to progressive concentration ratios (C/C_0) for linolenic and linoleic acids as time (t) proceeds, as illustrated in Fig. 4.5.

Figure 4.5 shows that, as hydrogenation time progresses, the compositions change. Although the curve for linoleic acid is rather flat, there is a predictable time for its maximum value.

The shape of the curve can be altered, because of the different activation energies, by shifting the temperature of operation.

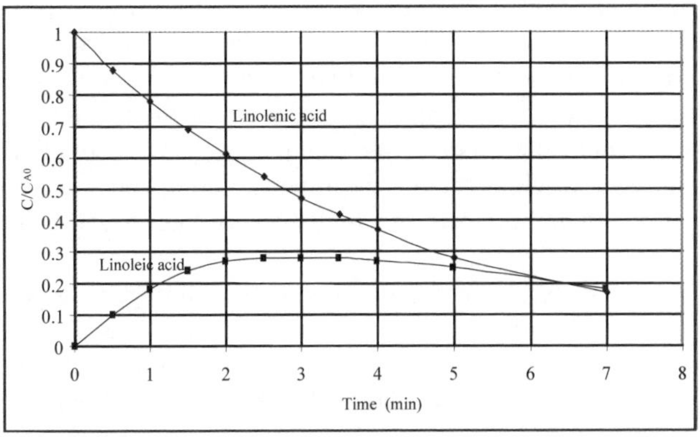

Data from Chen *et al.* (15)

Fig. 4.5. Progress of a hydrogenation reaction

The investigations necessary to set up and use such analysis are extensive, and may be worthwhile only in a large-scale higher-technology context such as fat hydrogenation. But the general pattern and nature of the chain reaction, and the implications, can be helpfully considered in quite a number of chain reaction systems that are encountered widely in food processing.

4.4 Parallel Sets of Reactions

Most foods are complex mixtures of constituents. The processing conditions, such as temperature, are necessarily applied to all the components. Sometimes, components can be easily isolated and processed separately, but, under other circumstances, this is not practicable and the food, or appreciable portions of it containing a number of constituents, has to be processed as a whole. So there are a number of reactions occurring in parallel; e.g., in bread baking, there are sugars going to carbon dioxide and alcohol, starch gelatinisation, protein denaturation, and caramelisation of sugars. On heating the bread, all reactions are speeded up, and these increased changes will affect the final product attributes.

A further complication is that, in many heating situations, for example those involving conduction such as baking, regions of the food remote from the heat source rise in temperature much more slowly than those close, so impacts of processing differ through the food.

So the situation commonly arises of sets of potential reactions implicit in the nature and composition of the food, starting up and proceeding, each at different rates and with often quite different degrees of desirability so far as the ultimate product is concerned. Therefore, the processor should consider all the possibilities and then optimise the processing conditions over these sets of parallel and linked reactions to arrive at the best food product outcome.

One example of this has already been encountered when considering the F values throughout a can containing a convection-heated product in an experimental retort, shown in Fig. 3.4. Here the differences in the extent of processing for the critical reaction, food poisoning spore destruction, in selected regions of the can are at once apparent. By inserting the kinetic parameters for other reactions within the product, for example enzyme destruction and starch gelatinisation, the relativities for these can be explored; and so on for any other reactions that are of importance and for which the necessary kinetic data are known.

Now each set of different processing conditions will in general give a different pattern of product composition and attributes – a different final product profile. So the problem for the processor is to assess the relative value of each product profile. Then potential processes can be explored with a view to optimising desired product attributes.

In some cases, it may be adequate to consider averaging – probably best done on a volumetric basis and effectively equivalent to finding the hypothetical concentration that there would be in the completely mixed product. In other cases, there may be absolute limits, upper or lower, such as, for example, with critical spore-forming bacteria, where a prescribed maximum spore concentration may be laid down in regulations.

One way in which to visualise these reaction patterns is to make use of the OTT chart. For example, if a critical requirement is overriding, then this will specify a line on the diagram "below" which processing cannot be considered complete. All regions in which the extent of processing for the critical constituent is equal to or greater than the level defined by a line on the diagram will lie on or above this line. Processes corresponding to all points on the line will have the same critical processing, so they will meet this requirement. But these processes may, and generally will, have differing extents of other parallel reactions, and therefore result in different product profiles. So the OTT chart can be used as the basis for exploring the possibilities and judging between them. In Example 4.5, the use of OTT charts in selecting times and temperatures for the pasteurisation of milk is described. These approaches are discussed further in a number of accounts, including Kessler (7) and Lewis & Heppell (8).

Example 4.5: Pasteurisation of milk

An important industrial process is the pasteurisation of milk, whereby the milk is heated for a period principally to destroy possible pathogenic microorganisms.

Taking one processing specification of 15 s at 72 °C and assuming a z value of 8 °C, then an OTT chart can readily be constructed for this, and if desired extended to include under- and over-pasteurisation expressed for example as percentages.

Alternatively, if it is assumed that at 72 °C the 1D time of *Mycobacterium tuberculosis,* the original organism for which the treatment was designed, is about 1 s, then lines of equal effect can be plotted. To indicate levels of destruction, three lines are shown 12D and 18D, as well as the 15D line for "pasteurisation" in Fig. 4.6.

Tests have evolved for adequacy of pasteurisation using chemical assays, and one of these is to test for alkaline phosphatase. If it is assumed that the test registers a level of 95% destruction of alkaline phosphatase, reached in about 20 s at 72 °C, and the z value for the phosphatase is about 7 °C, then a line for the destruction of this enzyme can also be added to the OTT chart.

All of these lines are shown in Fig. 4.6.

This OTT chart shows that, if 95% destruction of phosphatase is measured and therefore has occurred, then the processing required for 15D destruction of the critical bacteria has also occurred (and also that, of course, for 12D destruction, but not necessarily that for 18D).

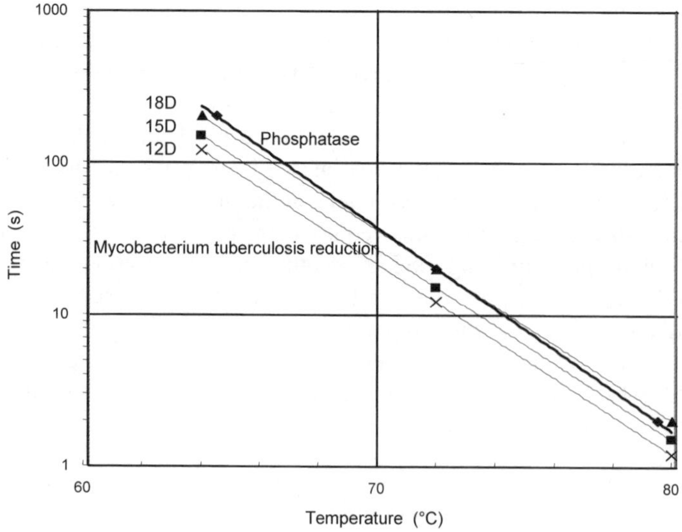

Data from Kessler (7)

Fig. 4.6. Pasteurisation of milk – OTT chart

If sufficient information is available, then the important reactions encountered in the processing of a single food can be explored kinetically, and the results of these experiments entered onto a single diagram. This then allows the overall processing situation to be assessed by the technologist. Obviously, critical requirements must be met. If there is more than one of these, then that with the most stringent demands overrides, and the OTT chart provides a minimum process line, on or above which any selected process must lie. However, this still allows a complete region for allowable processes, and just where in this region the actual process conditions will be selected becomes a matter of priorities and judgement.

> *Think break*
>
> * Can you think of other food processing examples in which the kinetics are well enough understood to set up an OTT chart for the significant reactions?
>
> * If you have difficulty in finding such examples, why do you think this might be so?

Data have been published that provide sufficient detail on the high-temperature processing of milk, to be used for illustration of optimising a set of reactions, and are studied in Example 4.6.

Example 4.6: High-temperature processing of milk

A number of reactions have been explored in the high-temperature processing of milk, and from these an OTT chart can be constructed, which is shown in Fig. 4.7. These reactions are extensively discussed in Kessler (7).

This OTT chart includes:

- bacterial reduction, looking at both thermophilic and mesophilic sporeformers. It can be seen that the activation energies (z values) are equal but, as would be expected, the thermophilic spore-formers can tolerate higher temperatures. These lines can be regarded as typical rather than specific; with further specific data, lines that are important in any particular situation can be substituted.
- destruction lines for constituents that have nutritional significance, such as lysine and thiamin, at two destruction levels, and destruction of 90% of a protease enzyme. Assuming kinetics for these, other lines can also easily be inserted to demonstrate the nutrients reaching other levels of destruction.
- a line indicating a panel judgement of the level at which a significant detectable change occurs in acceptability criteria, in this case colour and a cooked flavour, which emerge essentially simultaneously.

To indicate how such a chart might be used in practice, assume a critical requirement is that the mesophilic spores must be reduced by 10^9. Also, the protease destruction is important to the 90% level. Therefore, the allowable processes must lie in the regions on or "above" lines H and D. Taking flavour also into account limits selection to the region bounded by lines H, D and E.

This then restricts the possibilities to quite a small region on this particular diagram, and for example it can be seen that a process of about 5 min at 120 °C meets the requirements. Any further exploration of feasible conditions can then roam over the allowable region to see whether other criteria, such as, for example, minimising the residual lipase, line C, also needs to be considered. The effects of such a process on the thiamin and lysine can also be interpolated by inspection.

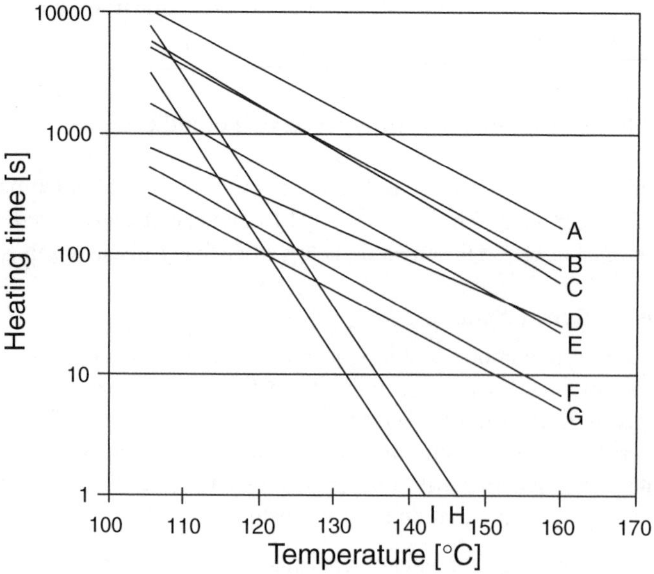

Destruction of **A** 50% thiamin; **B** 10% lysine; **C** 90% lipase; **D** 90% protease.
Just noticeable appearance of colour and cooked flavour, **E**.
Destruction of **F** 1% lysine; **G** 3% thiamine.
Reduction by 10^9 of **H** thermophilic spores; **I** mesophilic spores

Data from Kessler (7)

Fig. 4.7. High-temperature processing of liquid milk OTT chart

The OTT charts are constructed for a constant temperature process, and, of course, this is seldom if ever totally true in practice. However, quite a number of real situations are reasonable approximations, such as those encountered in many continuous heat exchanger situations. In some other real processes, sufficient of the bulk of the reaction occurs at the "working temperature" for constant temperature to be a reasonable approximation overall. Where this is not the case, then the OTT chart can still be used, reading off the reaction rate constants corresponding to actual temperatures and inserting these into the kinetic integrations. In other cases, processes can be divided into different segments, and each segment treated as a separate process. This is straightforward in the case of zero order reactions using simple proportioning. This can also be used as an approximation for reactions of other orders that is often compatible with what may be only quite moderate accuracy of the available data, and sufficiently close for the needs of practice.

4.5 More Complex Situations

Normally, the OTT chart giving lines of equal extents of processing shows a family of parallel lines, because, for a single reaction, the activation energies, which determine the line slopes, are constant irrespective of the stage of the reaction. This is illustrated for example in Figs 4.6 and 4.7. The families of lines can easily be extended if the reaction orders are known.

However, for some systems, the experimental lines when measured are not parallel, and an illustrative set of these for the precipitation of whey proteins from milk, derived from data given by Agrawala & Reuter (9) is given in Fig. 4.8.

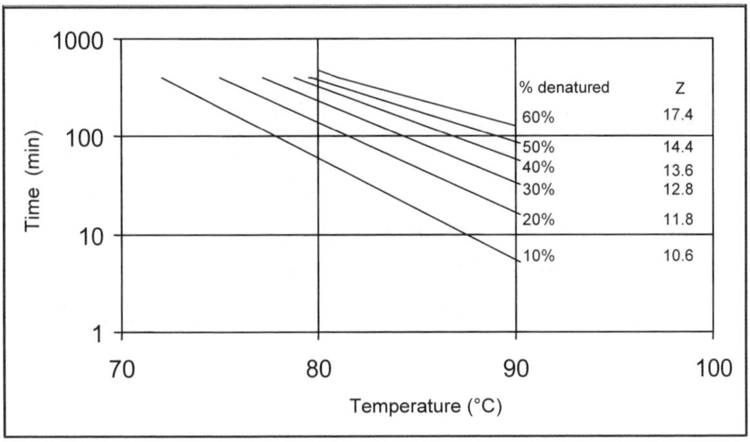

Data from Agrawala & Reuter (9)

Fig. 4.8. Precipitation of whey proteins: OTT chart

In Fig. 4.8, process lines of equal percentage of proteins precipitated are shown in the usual way, and it can be seen visually and from the tabulated equivalent z values that they are not parallel but systematically decrease in steepness as the fraction precipitated increases. This suggests that the residual proteins, as parts are removed by precipitation, change composition significantly. The residual proteins have progressively lower activation energy, thus contributing to a lesser slope of the curve.

Thus it can be seen that the OTT charts can provide much useful information. This can be both of direct and of inferential use to the processor. The major problem is often obtaining the data necessary for them to be constructed. It will have been noticed that many of the cited examples are of milk products, and the major reason for this is that milk systems have been both subjected to extensive scrutiny and widely published by the dairy industry. With milk so widely consumed and the dairy industry so extensively organised, it is not surprising that so much detail is available publicly about its kinetic properties.

Another feature emerging from experimental plots of d(lnk) against 1/T is the existence of broken Arrhenius curves for milk protein heat denaturation. This probably indicates a change in the reaction mechanism with temperature. When there are two pathways available, with different activation energies, then their relative significance changes with temperature. The pathway with higher activation energy "takes over" from that with the lower activation energy at the appropriate temperatures. This is illustrated in Fig. 4.9, from the work of Dannenberg (10). From 70-80 °C, the denaturation of α-lactalbumin has an activation energy of 268.6 kJ/mol; and from 85-150 °C, an activation energy of 69.01 kJ/mol.

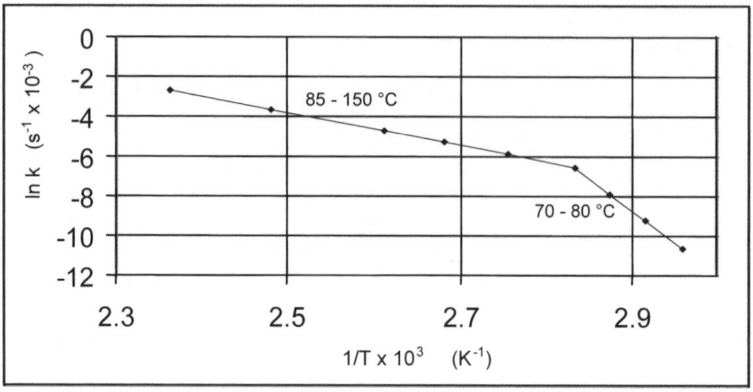

Data from Dannenberg (10)

Fig. 4.9. Denaturation of α-lactalbumin: broken Arrhenius plot

Example 4.7: Denaturation of α-lactalbumin

If the denaturation of α-lactalbumin in milk is a first order reaction, as suggested by Dannenberg (10) and the rate constant is given by:

$k = 10^{36.87}\ exp(-268,600\ /RT)$ for temperatures between 70 and 80 °C
and
$k = 10^{6.93}\ exp(-69,010\ /\ RT)$ for temperatures between 85 and 150 °C
where k is measured in (s^{-1})

What is the expected time needed for the 90% denaturation of this protein constituent of milk at temperatures of 75 °C and 100 °C

At 75 °C $k = 10^{36.87}$ exp(- 268,600/RT) = exp(- 268,600 / {8.314 x348)
= exp(-7.924)
= $3.62 \times 10^{-4}\ s^{-1}$
$-\ln C/C_0 = -\ln 0.1 = 2.303 = kt$

and so time for 90% denaturation = $\mathbf{t_{0.9} = 2.303/k = 2.303/3.62x10^{-4}}$ **= 6,363 s = 106 min**

At 100 °C $k = 10^{6.93}$.exp (-69,010/R x 373) = $1.85 \times 10^{-3}\ s^{-1}$
and again
$-\ln C/C_0 = -\ln 0.1 = 2.303 = kt$

and so time for 90% denaturation = $\mathbf{t_{0.9} = 2.303/k = 2.303/1.85\ x10^{-3}}$ **= 1,245 s = 21 min**

The broken curves can be handled by treating the straight line portions separately but using the standard methods.

The behaviour of the denaturation/temperature curves in Example 4.7 could be explained by postulating two consecutive reactions. The first reaction, dominant at high temperatures and having low activation energy, and the second dominant at the lower temperatures and having high activation energy, but more evidence would be needed to establish this.

4.6 Process Optimisation

In practical process situations, the temperature histories throughout the mass of the food are physically linked. This is obviously and inevitably true, for example, in a conduction heating system. In this, it takes time for changes in temperatures to penetrate through from the heat transfer interface, which could be fat in a frying system, into the more remote regions of the food, perhaps into an emulsified sausage that is being fried. It can be assumed that the various constituents of the food are evenly mixed at any one point, that is the food is homogeneous. Then the temperature histories of all the different constituents will be the same. But, because of the different reaction rate constants, the reaction patterns will not be the same in all, or perhaps indeed any, of these constituents.

The set of time/temperature conditions throughout the food can be thought of as a "process envelope", of identical temperature histories, but leading to different outcomes for the different ingredients. The processors usually have direct control over the temperature and time conditions; for example, the heating medium temperature can be altered. They have some indirect control – for example by modifying flow patterns to change the heat transfer coefficients at the surfaces. They have some possible control by changing the size or the composition of the food pieces (but this also changes the product). But, over thermal conductivity and activation energy, they have no control, as they are determined by the ingredients used and their constituents. Of course, the ingredients can be replaced with substitute ingredients, which may have constituents with different physical properties, for example replacing native starch with treated starch, or milk protein with soya protein.

Think break

In a typical food process based on heating:

* List the actual process controls available to the processor.
* How might the extent, the precision, and the repeatability of these controls be modified? Improved?

Unfortunately, because of the non-linear nature of most of the reaction processes and of the temperature coefficients of the reaction rate constants, the process outcomes from multiple reactions, even when the whole set is subject to the same overall temperature, are complex and analysis is not straightforward. However, by using the methods set out, so long as the kinetic behaviour of the critical and important reactions are known, outcomes can be predicted for any given process envelope. Then these can be re-evaluated for any other process envelope. So, with the aid of computers and spreadsheets, outcomes can be explored without too much labour. In particular, it should be possible to reveal overall trends in the significant constituents. These are generally sufficient to show the areas where actual processes will be both operationally practicable and straightforward, and also yield something close to optimum product outcomes.

Much time and energy have been expended over the years, and at an accelerating rate with the increasing exploration of new techniques, on seeking optimum processes. A wide-ranging but brief summary of many of these techniques was given by van Loey *et al.* (11). Another account, also with references and some detail, is in Holdsworth (12). Because most of the processes are complex, the procedures tend towards searches of possible alternative heating strategies, which are then explored for their relative effects on critical and important variables. Amongst the published material are accounts of both fixed temperature and variable temperature strategies. If process variables other than temperature were driving the reaction, the same general methods could be used.

For example, an early paper by Teixeira *et al.* (13) illustrated the complexity of optimising the product profile from conduction-heated systems in can sterilisation. Safety is, of course, critical and there has to be a fixed least-processed-point lethality (F_0). In considering the other product attributes, the normal inference is that higher temperatures and shorter times favour overall quality but the Teixeira paper (13) showed that this was true only up to a point so indicating an optimum process. In these heat-conduction packs, it was the relative importance of the heating and cooling rates in the different regions going down to the can centre that affected the average levels of the other product attributes. This led towards what amounted to an optimum process condition for such solid packs.

Another examination of conduction systems by Silva *et al.* (14) looked systematically at both average pack quality and also surface quality, and at the relative effects of sterilisation temperatures on these. Durance *et al.* (15) examined variable retort temperatures for the canning of salmon to see whether quality could be improved or process times decreased, within constraints practical in cannery operation. They treated it as a conduction-heated pack, and looked at surface "cook" effects and nutrient retention while maintaining constant centre point lethality (F_0). They used relatively straightforward mathematical techniques. They concluded that programming variable retort temperatures offered advantages over fixed ones, although the differences were not major.

Optimal heating strategies for a convection oven, considering the problems of re-heating pre-prepared meals but also more generally applicable, were discussed by Stigter *et al.* (16) using quite elaborate mathematical procedures. They were looking particularly at control and regulator design for the ovens, and achieved their objective of achieving uniform temperatures within the food at the end of the heating time. The examples briefly outlined seem to offer possibilities and ideas for controlling processing reactions.

These few case studies indicate something of the extent and scope of optimisation investigations in food processing. They demonstrate how reaction technology, together with the necessary accompanying experimental data, can be used to seek improved conditions for working processes across a very wide range of industrial situations. There are several research groups, internationally, that are working hard and effectively in this field. The academic groups, especially, are generally publishing their results. Examination of the current literature, both the published papers and the extensive bibliographies cited in many of them, quite rapidly focuses on the information that has a bearing on any particular processing problem, and provides important indicators towards investigation, selection and operation of better processes.

Think break

To find recent information on kinetics of reactions, search the last 2 years of the Food Science and Technology Abstracts for:

* shelf-life of whole milk powder
* blanching of vegetables
* bread baking

Case study 4: Heat treatment of milk

This case study shows how, when there are several critical and important reactions, an optimum process can be designed.

The various reactions in milk processing have been assembled in a study by Arteaga *et al.* (17), as shown in Table 4.I.

TABLE 4.I
Kinetic data for reactions in the heat treatment of milk

Reaction (order)	k_{120}	E kJ mol^{-1}	ln A s^{-1}
Lipase inactivation (1)	0.0115 s^{-1}	53.037	11.767
Protease inactivation (1)	0.0125 s^{-1}	63.963	15.194
Furosine formation (0)	0.497 μmol $l^{-1}s^{-1}$	81.637	24.286
Lysinoalanine formation (0)	3.87×10^{-2} μmol $l^{-1}s^{-1}$	101.377	27.775
Lactulose formation (0)	5.38 μmol $l^{-1}s^{-1}$	120.224	38.478
Thiamin loss (2)	7.53×10^{-4} l $mg^{-1}s^{-1}$	100.800	22.742
Lysine loss (2)	2.12×10^{-8} l $mg^{-1}s^{-1}$	100.900	15.691
Colour formation (1)	1.66×10^{-3} s^{-1}	116.000	29.101
HMF formation (0)	0.22 μmol $l^{-1}s^{-1}$	135.098	39.833
Micrococcacea destruction (1)	1.29×10^{-5} s^{-1}	329.985	112.759
B. stearothermophilus spore destruction (1)	1.10×10^{-2} s^{-1}	345.357	101.188

k: reaction rate constant, E: activation energy, A: frequency factor
Data cited by Arteaga *et al.* (17)

Contd..

Case study 4 (contd)

From Table 4.1, OTT charts can be drawn. However Arteaga *et al.* (17) took a somewhat different, a numerical, approach.

They postulated what they called a *four-step process*, where the UHT heating process was divided into four time/temperature steps. Each step was assumed to be at a constant temperature, which were rising and then falling, for example sequential temperature steps at 80, 120, 140 and 100 °C.

They then assumed some critical process outcome requirements, that is levels of product attributes, which they termed constraints. For example, in one of their illustrative systems, they required that at least 40% of the protease and lipase activity be destroyed, while minimising the increase in the colour and hydroxymethyl furfural (HMF) formation, which is an indicator of the browning reaction.

They then used an interactive procedure using a constrained simplex ("Complex") method, starting from an initially arbitrarily assumed four-step process with specific time/temperature limits for each step. The computer program included a kinetic subroutine, which was based on the kinetics constants according to the Arrhenius equations. For each of the reactions, the total heat effect at the end of the time/temperature profile was taken as the sum of the effects at each step.

The Complex method, from the original process meeting the constraints, by reiteration developed new temperature profiles, maintaining the constraints on lipase and protease, but at the same time minimising the effects on colour and the HMF. This then generated systematically improved minimal colour and HMF, until an optimum was reached, whilst still meeting the constraints.

The optimum result they obtained is summarised in Table 4.II.

TABLE 4.II
Seeking optimum time/temperature profiles in four-step heat process

Initial time/temperature profile

Step 1 8.4 s @ 80 °C	**Step 2** 25 s @ 120 °C
Step 3 20 s @ 140 °C	**Step 4** 19.5 s @ 100 °C

Process result: HMF 20.66, colour 0.16, residual lipase 40.2%, residual protease 35.5%.

Optimum time/temperature profile

Step 1 20 s @ 79.3 °C	**Step 2** 21.1 s @ 115.7 °C
Step 3 6.9 s @ 139.8 °C	**Step 4** 19.6 s @ 99.8 °C

Process result: HMF 8.05, colour 0.07, residual lipase 59.97%, residual protease 58.33%.

Units: HMF mg ml^{-1}, colour in arbitrary units. Lipase and protease are residual activity as % of original activity. These were constraints that had to be less than 60% residual activity.

Contd..

Case study 4 (contd)

This result minimises the heat damage effects, HMF down from 20 to 8, and colour change about halved, while at the same time meeting the required level of enzyme destruction (at least 40% destroyed).

Constraints on other product attributes can be set either as stipulated constraints or minima. The final values for all the product attributes with the optimum time/temperature process are also calculated so that effects on the other attributes as well as the constrained and minimised attributes are known.

While the four-step process may seem rather arbitrary, it is a reasonable compromise in terms of complexity and it indicates the manner in which such procedures can be employed. Higher levels of complexity could be taken into account although they may not yield substantial improvement in the total process. While the method itself does not necessarily picture the field of possibilities all together, its use would also serve to generate familiarity and understanding of the reaction system and build a basis for process design.

Once again, the method depends upon sufficient knowledge of the kinetics of the important reactions, and the effectiveness of such an approach obviously stands or falls by the adequacy of this knowledge.

Because the OTT chart is helpful and instructive, it is tempting to invest it with wider strict application than it actually has. Recall, from its derivation, the OTT chart applies only to what can be called isothermal processes. These are processes where the temperature is lifted from a lower level, at which effectively no reaction occurs, suddenly up to one at which there is substantial reaction. The temperature remains at this elevated constant value for some time, during which the reaction progresses to the controlled extent, and thereafter is instantaneously lowered again to a temperature at which reaction effectively ceases. Strictly, such a process cannot occur in practice, but close approximations can, such as in continuous heat exchangers, steam jet heating and vacuum cooling. Other practical processes sometimes come close enough for the analysis to have application. Thus the OTT chart provides the technologist with valuable and perhaps otherwise unobtainable insights, such as appear to have been important in developing a liquid egg pasteurisation process.

Case study 5: Pasteurisation of liquid whole eggs

This case study shows how knowledge of reaction technology can be used to design a process that can be protected by a patent.

Adding storage life to food ingredients adds important flexibility, and therefore processes that can increase high quality holding times are important and valuable.

Eggs are a widely used food ingredient, and qualities that have to be retained during holding include functionalities – for example the ability of the egg proteins to remain soluble and to whip up into typical egg products such as batters and cakes. Also, eggs may become contaminated by pathogens so that some degree of heat pasteurisation may be highly desirable. If the heating is sufficient to remove health risks, then it can easily coagulate the protein and destroy functionalities (Hou *et al.* (18)). So this is an area in which it might be possible to organise the heat processing to reduce possible pathogens to safe levels and at the same time preserve protein structures, by knowledge of the kinetics of the important reactions and then application of reaction technology.

A practical method of doing this in continuous flow, high temperature, short time pasteurisation equipment, is claimed in a US patent (19). The patent includes kinetic data, and an OTT chart, which is used to explain some of the basis for the claims. Figure 4.10 is a redrawn OTT chart, taken from equations included in the patent descriptions.

Figure 4.10 includes a destruction line for *Salmonella* spp, which are well known to be problem pathogens in eggs, and for which the 9D reduction conditions as shown may be regarded as a target reduction ratio. It includes a 7D reduction line for another common pathogen found in eggs, *Streptococcus faecalis*.

It also includes a line indicating the heat treatment that results in 5% loss of the soluble egg proteins and which may be seen as an acceptable price, in terms of degree of loss, to pay for much enhanced product safety. The 5% loss of soluble protein line shown is a broken one, indicating something of the complexity of the proteins, and was determined by detailed experimental studies (18).

In the upper right-hand corner of Fig. 4.10 is a region of much increased protein coagulation within which the eggs become functionally inadequate for many ingredient applications. Thus an OTT diagram was used as background to the design of novel processing, described in the patent, and claimed to make liquid whole egg products with pre-selected, extended, refrigerated shelf-lives.

Contd..

Case study 5 (contd)

This patent is instructive, entirely apart from any intrinsic merit it may have, because it displays a considerable kinetic background to the product claims, and apparently regards this as important to the invention. This seems to imply that the process might not have been discovered without the kinetic analysis. It also shows how the kinetic information relates to the advantages claimed to derive from use of the patented process, and which are exemplified by product applications such as sponge cakes and custards.

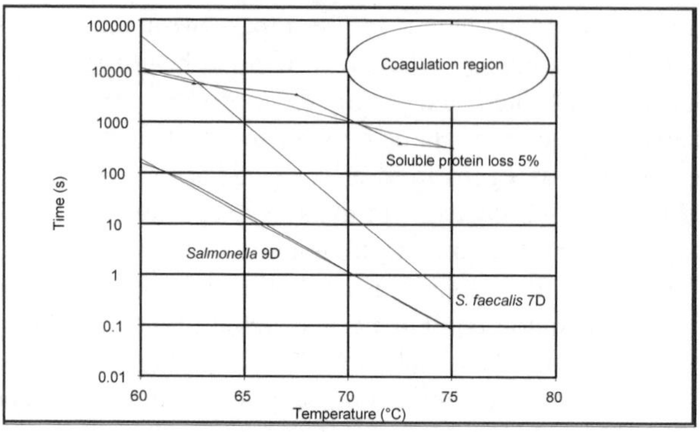

Data from Swartzel *et al.* (19)

Fig. 4.10. Pasteurisation of whole egg – OTT chart

4.7 Processing in Continuous Systems

Some foods are processed continuously. The continuous flow processing of foods that can be pumped introduces a new constraint imposed by the physical situation, residence time distributions. For example in pipe flow, elements on the centreline move faster and therefore spend less time in the processing than elements on the periphery, and this is inevitable from the flow patterns. Therefore, they will be much less processed and so will affect the overall quality assumed to be an average over the whole flow. A distribution of residence times is needed to cover all the elements. Analysis of flow systems is complicated but some of the features encountered are illustrated in a Case Study of an industrial food process that uses a continuously stirred tank reactor. In this, the reactor is a large tank maintained at the process temperature. The food is pumped in continuously, immediately heated and uniformly mixed, and an average sample from the whole tank flows out at the same rate as the inflow to keep the tank contents volume constant. This system achieves very rapid, effectively instantaneous, heating of the food, but it also inevitably incorporates an exponential spread of food residence times in the tank.

Case study 6: Continuous heat processing of tomato paste

An example of the use of reaction technology analysis in continuous processing has been investigated in an industrial application, the processing of tomato paste.

To provide adequate keeping properties for concentrate for sauces, it is important to heat-process the pulped tomato, but this heating has, as far as possible, to avoid breakdown of the pectins that give the resulting paste its structure. Hydrolysis causing breakdown of the pectins is enzymically catalysed by pectinase (polygalacturonase) and is quite strongly heat-dependent. So to minimise hydrolysis the processes can destroy the enzymes by moving quickly to higher temperatures above 50-60 °C, at which temperatures the enzyme activity effectively ceases. A further complication arises from differential reaction rates in tomatoes of different ripenesses, input tomatoes to a plant being not always equally ripe.

In a recent account of experimental and theoretical work by Srichantra (20), activation energies of the various potential reactions have been related to the rates and temperatures of the reactions in the processing. The work shows that reduction in pectin concentration, as measured by galacturonic acid, which is a breakdown product of the pectin indicating structural loss, can be predicted and quantified for various possible conditions of processing.

To obtain rapid heating, a large heated "break" tank was employed, with pulp continuously pumped in and removed. To analyse the situation, Srichantra inserted appropriate and measured values for tank volumes, flow, reaction rates including those for tomatoes at different ripenesses, and activation energies, into differential equations. These equations were then solved for selected tank temperatures by using computer algorithms. The results demonstrated the effects of both various tank temperatures and nominal paste residence times, and were extended to tomatoes of differing ripeness. The method could be further adapted to other processing circumstances.

Particular effects of tank temperature, paste residence time, and tomato ripeness are illustrated in Figs 4.11(a) and 4.11(b), clearly demonstrating the peaking pectin loss at temperatures around 60 °C. This effect, leading to the cold break (temperatures <60 °C) and hot break (temperatures >60 °C) processes for tomato paste, is well known, but the analysis

Contd..

<u>Case study 6 (contd)</u>

explains how it results from the interaction of the rates of concurrent reactions – enzymic breakdown of pectin and enzyme denaturation. The advantage of the analytical approach is to quantify the behaviour and so provide data for more precise operation of existing processes as well as indicate possibilities for the design of new ones.

This general behaviour involving enzyme activation/inactivation by heating is encountered often in fruit and vegetable processing so the analysis has wider applicability. It demonstrates that quite sharp process changes can occur and so the labour of experiment and analysis to identify these quite precisely may well be worthwhile. The model also shows quantitatively how both changes in factors such as fruit ripeness leading to differing rate constants, and also rapid heating that can be obtained by pumping directly into hot material, can be used in process design to explore quality outcomes.

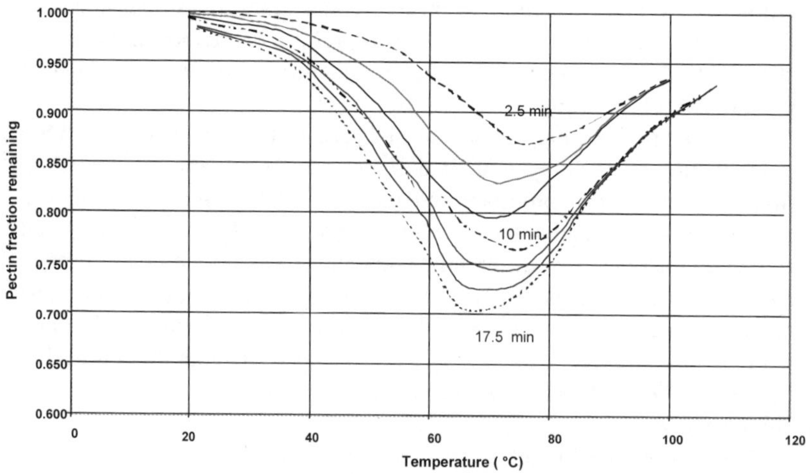

Fig. 4.11(a). Pectin remaining as a function of residence time and break temperature

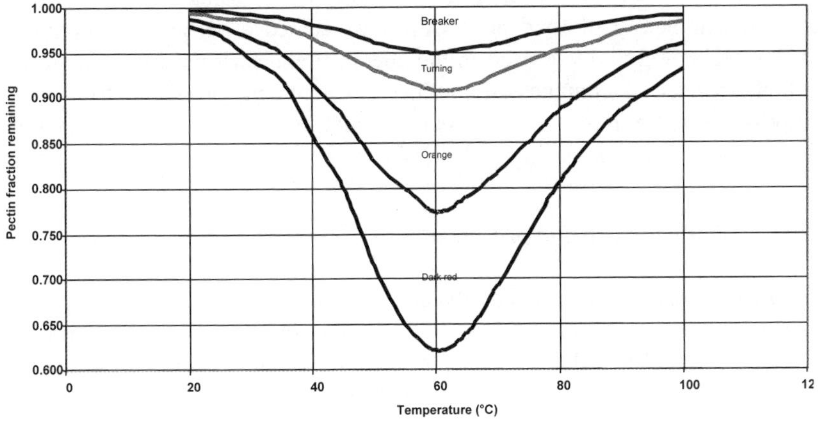

Fig. 4.11(b). Pectin remaining as a function of ripeness and break temperature

4.8 Practicalities

The methods of reaction technology can be applied in all food processing situations in which the food constituents undergo the reactions that change them and therefore alter quality attributes in the food. So they are very practical, and it is worthwhile to briefly summarise some of the important practical approaches that have emerged. These may also indicate the scope that undoubtedly exists for further analyses along the same lines, but extending beyond what has been accomplished already.

4.8.1 Examining constituents and attributes

The last four chapters have emphasised that in every process there are raw materials that change in the process to give the final product. This final product can be characterised by critical and important attributes, and the aim of the processing is to achieve the specified levels of these attributes. These attributes can be identified with specific reactions occurring in the process. The reactions are controlled by the processing conditions, and in particular temperature, time and concentration have been studied in detail in the previous chapters. The OTT charts and the related reaction rate constant calculations can be used to determine the effect of time, temperature and concentration on the extent of individual reactions. For example, the OTT plots can be compared under different conditions, and the critical conditions can be identified, as can also the processing "area" for optimum levels of the important product attributes.

This gives the basis for either improving a present process or designing a new one. More generally, for the future it leads food processing into an expanding area of knowledge, developing it as a more advanced processing technology. In the past, the complexity of food processing has seemed a barrier to applying the reaction technology widely used in the chemical and pharmaceutical industries. With the modern techniques of product development, product attributes and compositions are well recognised in the food industry, and the reactions leading to them are now being increasingly identified.

4.8.2 Improving existing processes

Processes are usually specified by the critical product attribute, very often safety. In the past, there was optimisation of food processing, but it was done as a craft by craftspeople. When the craftspeople were highly competent, the results were very good; but they were not able to take advantage of the systematic methods of modern technology. Therefore, their results were limited by the knowledge and instrumentation available; they were not always reproducible, and not easily transferred to new operators and operations, or to automatic control equipment. Much advanced optimisation is now practicable.

Various processing regimes can be compared for their effects on the reaction rates and therefore the levels of the product attributes. The dairy industry has recognised this over many years and has developed new processes such as high temperature/short time (HTST) heating. They have built, from large-scale research in the laboratory and in the plants, knowledge of reactions and the effects of the processing conditions on the reactions. Information technology has given them the tools to relate the levels of product attributes (from quality assurance records) to the processing conditions (from the plant records). From this and the laboratory research, they have built models of the process. These models can then be used for process control to improve the effectiveness and the efficiency of the process. Another food industry developing these approaches to processing is the Australian and New Zealand wine industry, which has moved from a craft to large-scale technological industries.

Some food industries may think that, for them, it is too difficult; they have multiple raw materials, and are producing a wide variety of the products on the same plant. But they can also analyse what are the fundamental product attributes in all the products and the reactions causing them. This may simplify what they think is a confused situation, and so lead to more systematic processing. They can examine the possible variations of the process conditions, and their effects on the reactions related to the fundamental product attributes.

Perhaps the greatest use has been in the prediction and control of the shelf-lives of foods during distribution and marketing. Because the times are comparatively long and the reaction rates are usually slow, predictions can be well within the accuracy of the distribution control. As has been seen in the examples in Chapters 3 and 4, shelf-lives can be predicted for storage at all temperatures, from frozen storage to tropical conditions. Therefore, reaction technology has proved a useful tool in design and control of distribution systems. Computer software programs based on such predictions have been developed and are available (21).

Processing possibilities, such as speed of heating, accuracy and uniformity of process control, can be studied using OTT charts, and Arrhenius plots. Reaction rate constants and sensitivities can develop a "feel" of the process as regards time and temperature.

4.8.3 Application to new product design and process development

The recognition of the inter-relationships between product design and process development in some companies has led to some fruitful innovations. Food product design came from the craft base of the chef and the small food processor, of empirical (although knowledgeable) try and taste. But, in recent years, the technological knowledge has grown. There is an ever-increasing knowledge of product attributes and the methods of measuring them – chemical, physical, biological and sensory. Perhaps the greatest advances have been in sensory science, with the increasing identification of attributes, quantitative methods of measuring them and correlation with physical methods.

This has led to the use of experimental designs to study the effects of different levels of ingredients/raw materials and processing conditions on the individual product attributes, and the optimising of the important product attributes. By use of the OTT charts for different product attributes, the "ball park" areas for time and temperature can be identified, as described in the section on multiple reactions, and these can be used as the basis for the experimental designs. Knowledge grows on the effects of different raw materials and other process conditions, such as pH, humidity, pressure and controlled modified atmospheres. This can be used as the knowledge base for product development. Examples of new products, which have already come from knowledge of reaction technology, are the various milk protein fractions that are manufactured for a wide variety of food products. These ideas will be examined further in the final chapter.

Think break

In Think breaks in Chapters 1 and 2, you studied four food processes, identifying the critical and the important product attributes and the related reactions. Choose one of these food processes, identify the "envelope" of the process and study how you could optimise the process.

* Identify and consider the present and possible processing conditions.
* If available, study the OTT charts for the critical and important reactions. If not, use the data from the last three chapters as "general" reaction kinetics for comparison of the temperature and times for the different reactions.

Contd..

Think break (contd)

- Identify the process envelope for the times and temperatures.
- Discuss how optimisation procedures might be applied to the possible processing conditions to secure the specified product.
- Design experimental procedures to study your suggested process(es).

4.9 References

1. Trezza T.A., Krochta J.M. Color stability of edible coatings during prolonged storage. *Journal of Food Science*, 2000, 65, 1166.
2. Wang R., Schroeder S.A. The effect of caramel coloring on the multiple degradation pathways of aspartame. *Journal of Food Science,* 2000, 65, 1100.
3. Roy S.S., Taylor T.A., Kramer H.L. Textural and ultrastructural changes in carrot tissue as affected by blanching and freezing. *Journal of Food Science*, 2001, 66, 176.
4. Aiba S., Humphrey A.E., Millis N.F. *Biochemical Engineering.* New York. Academic Press, 1965.
5. Jousse E., Jongen T., Agterof W., Russell S., Braat P. Simplified kinetic scheme of flavour formation by the Maillard reaction. *Journal of Food Science*, 2002, 67, 2534.
6. Chen A.H., McIntyre D.D., Allen R.R. Modeling of reaction rate constants and selectivities in soyabean oil dehydrogenation. *Journal of the American Oil Chemists' Society*, 1981, 58, 816.
7. Kessler H-G. *Food Engineering and Dairy Technology.* Freising. Verlag A. Kessler, 1981.
8. Lewis M.J., Heppell N.J. *Continuous Thermal Processing of Foods.* Gaithersberg M.D. Aspen, 2000.
9. Agrawala S.P., Reuter H. Effects of different temperatures and holding times on whey protein denaturation in a UHT pilot-plant. *Milchwissenschaft,* 1979, 34, 735.
10. Dannenberg F. *Zur Reaktionskinetik der Molkproteindenaturierung und deren technologischer Bedeutung*. PhD Thesis, Technical University of Munich, 1986.
11. Van Loey A., Hendrickx M., Smout C., Haentjens T., Tobback P. Recent advances in process assessment and optimisation. *Meat Science,* 1996, 43, S81.
12. Holdsworth S.D. *Thermal Processing of Packaged Foods.* London. Blackie Academic and Professional, 1997.

13. Teixeira A.A., Dixon J.R., Zahradnik J.W., Zinsmeister G.E. Computer optimization of nutrient retention in the thermal processing of conduction-heated foods. *Food Technology*, 1969, 23, 845.

14. Silva C.L.M., Oliviera F.A.R., Pereira P.A.M. Optimum sterilization: a comparative study between average and surface quality. *Journal of Food Process Engineering,* 1994, 17, 155.

15. Durance T.D., Dou J., Mazza J. Selection of variable retort temperature processes for canned salmon. *Journal of Food Process Engineering,* 1997, 20, 65.

16. Stigter J.D., Scheerlinck N., Nicolai B., Van Empe J.F. Optimal heating strategies for a convection oven. *Journal of Food Engineering,* 2001, 48, 335.

17. Arteaga G.E., Vazquez-Arteaga M.C., Nakai S. Dynamic optimization of the heat treatment of milk. *Food Research International,* 1994, 27, 77.

18. Hou H., Singh R.K., Muriana P.M., Stadelman W.J. Pasteurization of intact shell eggs. *Food Microbiology,* 1996, 13, 93.

19. Swartzel K.R., Ball H.R., Mohammed-Hossain H-S. US Patent 4808425, 1989.

20. Mohammed-Hossain H-S. *Criteria Development for Extended Shelf-Life of Pasteurized Liquid Whole Egg.* PhD Thesis. North Carolina State University, 1984.

21. Srichantra A. Modelling of the Break Process to Improve Tomato Paste Production Quality. MTech Thesis, Massey University, 2002.

22. Blackburn C. de W. Modelling Shelf-life in the Stability and Shelf-life of Food. Edited by Kilcast D., Subramaniam P. Cambridge, Woodhead, 2000.

5. BROADENING THE NET

5.1 Introduction

So far, this book has mainly looked at thermal processing. The concentration of a critical or an important product component or attribute has been linked to processing time, through a reaction rate constant. And because so much of industrial food processing is initiated and controlled by changing the temperature, the effects of temperature on these reaction rate constants have been quite extensively reviewed. Sometimes, effects of time and temperature are sufficient in studying reactions in food processing, but at other times the effects of other processing conditions need to be considered, as illustrated in Fig. 5.1.

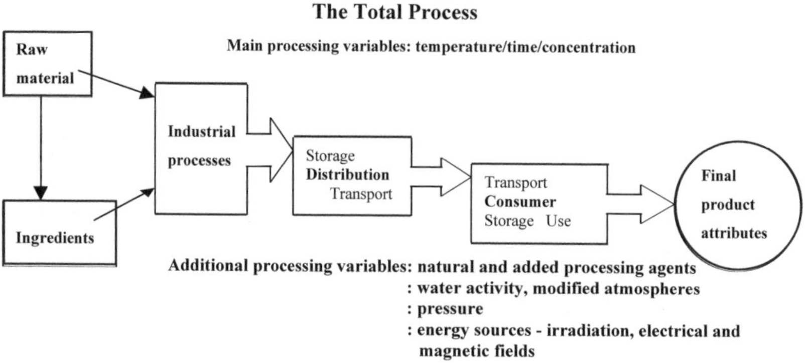

Fig. 5.1. Food processes and the processing variables

Figure 5.1 also emphasises that the "process" is not only the apparent industrial process but also the distribution and the consumer's handling of the food because it is the attributes of the final food as eaten that are important. The reactions and the resulting changes in the food materials during the whole sequence lead to the final product attributes. In foodservice, the control of the whole reaction sequence is technologically possible, but needs knowledge and technical skills. More commonly, food products are manufactured and distributed through complex chains to the final consumer, and it is difficult to have continuous control of the reactions.

Because of the effects on reaction rates of concentrations of the reacting materials and also of natural processing agents in the raw materials, purification and treatment of the raw materials to produce standardised food ingredients have become an important method of controlling the outcomes of the reactions. Processing agents can also be added to the food to increase or decrease the extents of the reactions.

Energy in forms other than temperature can, under suitable conditions, induce, hasten and slow food reactions. Irradiation, increased pressures, and electrical forces have all been suggested and sometimes used in food processing. They need to be taken into account, so that the technology of their use can be considered and their effects predicted. The fit of these into any general food processing pattern needs to be explored.

Water activity affects the rates of many reactions in which water is involved. Choice of the range of water activity limits the types of reaction that can take place; for example, lowering the water activity will firstly stop bacterial growth, then moulds and yeasts; specifying the water activity can be used to control the reaction rate.

Today there is increasing awareness of taking into account several process variables together, and designing the process so that optimum control of the critical and the important product attributes is achieved. Packaging design has greatly aided this, by being able to achieve a controlled atmosphere or vacuum, sterile filling and closing, pasteurising and sterilising of packs. The developments in controlling a number of process variables together have led to improved product attributes, but they also require considerable knowledge of the reactions and the effects of the different process variables when used together. In some cases, otherwise attractive new processes have been unsuccessful because of lack of understanding of the reaction technology.

5.2 Processing Agents

Because of the complex nature of foods, there are often components in the food materials that materially affect the critical and important reactions in the food; also, there may be processing agents that can be added to influence specific reactions. The concentrations of these other components must now be considered more carefully in the processing. Many common food constituents, such as water, acids and enzymes, can be regarded as processing agents as well as ingredients. Obviously, components in the food raw materials cannot just be added or subtracted at will as this may detrimentally affect the product. However, refining and purification to produce food ingredients can remove many of the interfering components; for example, by removing acids from fats by neutralisation, their effect on hydrolysis reactions can be removed. Processing agents can be added to increase or reduce the rate of the reactions – for example, sulfur dioxide to slow down browning and acetic acid to slow microbial growth. Even oxygen, in the atmosphere within which processing generally occurs, enters into many reactions as a significant reacting element.

5.2.1 Additives

Although there could be variations in initial concentration, this could often be only the biological variation in the raw materials, which may be quite minor. Concentration of natural processing agents in the raw materials, and added components from formulations can have a substantial effect on the reactions. So the concentration of added processing agents such as acids and alkalis, salt, sugar, glycerol, antibrowning agents, antioxidants, antistaling agents and essential oils has to be considered (1). There are many instances. An example is the influence of one group of additives, thermal protectants, on the heat denaturing of antibodies and antibacterial components in infant formulas, which has been investigated by Chen *et al.* (2). Table 5.I shows how the activation energies of the heat destruction of immunoglobulin G (IgG) changed with added processing agents, indicating the quite substantial effects of those thermal protective additives on properties that can be very important for processing.

TABLE 5.I
Activation energies of heat denaturation of immunoglobulin G (IgG) with thermal protectants

Sample	Activation energy (kJ/mol)
In phosphate buffer	
IgG in NaCl/phosphate buffer	328.4
IgG + 0.2% glutamic acid	300.5
IgG + 10% whole milk	289.4
IgG + 20% maltose	280.3
IgG + 20% glycerol	265.4
In colostral whey	
Whey	316.1
Whey + 0.2% glutamic acid	292.8
Whey + 20% maltose	273.6
Whey + 20% glycerol	257.3

Bovine milk IgG, in phosphate buffer (pH 7.0) and colostral whey, with added thermoprotectants, at temperatures 70, 72, 74, 78, 82 °C

From Chen, Tu & Chang (2)

These data show how addition of processing agents can change the activation energy of the heat denaturation and therefore both the rates and the relative rates of the reactions. The concentrations cited might not be suitable for some products, but the example shows how addition of thermal protectants can change the effect of temperature on the rates of reactions. Data such as these can be used in the planning of temperature regimes and heating and cooling rates, together with acceptable levels of processing agents, to achieve desirable final levels of the constituents in the final products.

Change of pH, increasing by adding alkalis and decreasing by adding acids, is a common processing strategy, for example in pickled vegetables and sauces to stop microbial growth. Acids affecting pH are often found naturally in food raw materials. Hydrogen ion concentration is normally quoted as the negative logarithm in gram mols per litre, the pH. For the first example of a rate equation in this book, the hydrolysis of sucrose, it was experimentally found that the rate was directly proportional to the concentration of hydrogen ions. This is also true of many other processing situations encountered in the food industry. An account of the effects of pH in different situations and processes is given by Tijskens & Biekman (3). They described simple equilibria and dissociation equations that they found to cover the discoloration in blanched vegetables and the activity of enzymes in different buffered and unbuffered systems.

Experimental results on the kinetic analysis of the breakdown of chlorophyll in green peas by Ryan-Stoneham & Tong (4) are given in Example 5.1. This loss of the bright green colour and development of a final yellowish green colour is common in canning green vegetables (4).

Example 5.1: Effects of pH on degradation of chlorophyll

In experiments on the thermal processing of pea purée, Ryan-Stoneham & Tong (4) examined the kinetics of chlorophyll (a) and (b) degradation as a function of pH. During the sterilisation process, there was a decrease in pH of the peas due to the formation of organic acids.

Rates of colour change and the effects on these of reaction temperature (80, 90, 100 °C) and pH (5.5, 6.2, 6.8, 7.5) were measured. In one set of experiments, the pH values were the starting pH and the pH was allowed to decrease naturally during the processing. They also measured the reaction rate under conditions of controlled pH (by the addition of gaseous ammonia to neutralise the increasing acidity).

Chlorophyll (a) and (b) degradation followed a first order model. With or without pH control, the temperature dependence always followed the Arrhenius relationship. They found their results were a good fit to an equation of the form

$$\ln k = \ln A - E/RT + c_p \, pH$$

where k is the reaction rate constant for the chlorophyll degradation, c_p is the coefficient describing the pH dependence.

Contd..

Example 5.1 (contd)

Using multiple linear regression on their data, the following equations were found to describe the reaction rate constants as a function of temperature and pH for both chlorophyll a and b:

Chlorophyll a $\ln k = 28.38 - 8796.2(1/T) - 1.193\text{pH}$
Chlorophyll b $\ln k = 25.53 - 8475.6(1/T) - 1.014\text{pH}$

The basic reaction was first order in both chlorophyll and hydrogen ion concentrations. That is, at constant temperature,

$$\text{rate r} = k\,(C_{Ch})\,(C_{H+})$$

where (C_{Ch}) is the concentration of the chlorophyll and (C_{H+}) the concentration of hydrogen ions and $-\log(C_{H+})$ is the pH.

Think break

In a company's tomato sauce, acetic acid is always present, and sometimes benzoic acid is added to control microbial growth. Marketing has decided that the acetic acid flavour is too strong, and benzoic acid is now not allowed in the Food Regulations.

* Discuss the microbial growth being controlled by these acids.
* List the other processing variables that could be used to control microbial growth.
* Select possible processing variables and their levels that could be used with acetic acid at lower levels to control the microbial growth.
* Describe the experimental studies you would conduct to find the processing conditions for the optimum product.

5.2.2 Modified atmospheres

Another common reactant in food systems is atmospheric air as the oxygen source for the oxidation of food components. Reducing the oxygen concentration in the ambient atmosphere can directly affect oxidation rates. The oxygen concentration can be controlled by reducing oxygen pressure inside packaging, introducing a nitrogen-enriched or carbon dioxide atmosphere, or conducting processing and packaging under a vacuum. This changes the partial pressure of the oxygen, which is proportional to its concentration. Modified-atmosphere packaging (MAP) has been a significant development in the last 20 years, combining gas atmosphere, temperature and time to ensure product safety as well as sensory attributes. For example, the storage life of fresh meat has been extended by providing much increased levels of carbon dioxide, which inhibits growth of microorganisms. Colour is then restored to the fresh red meat shade either when the package is opened to the air, or by having a packaging film that allows slow passage of atmospheric oxygen so that oxygen levels can slowly rise from, for example, an original 0.1% in the pack. Mathematical modelling has been used to follow such processing, the critical objective being safety. Data on the growth and survival rates of important microorganisms under different storage conditions (time, temperature, atmosphere, and also pH, water activity and meat composition) are the basis of the model. The program predicts the growth characteristics under particular conditions (5).

An early and continuing important use of controlled-atmosphere storage and modified-atmosphere packaging (MAP) is for extending the life of fresh fruit. The atmosphere controls the respiration of the fruit, and therefore first the fruit ripening and then the deterioration. A recent study on sweet cherries by Jaime *et al.* (6) studied the respiration rate for different levels of temperature, oxygen and time. Because respiration involves enzymic reactions, a basic enzymic rate model (Michaelis-Menten equation) was used, relating respiration rate and oxygen level. It was found that there were two different linear relationships, above and below 10±3% oxygen, following the linear relationship:

$$1/R_{02} = I/V_m + K_m / V_m [O_2]$$

where R_{02} is the respiration rate (mL(STP)kg^{-1} h^{-1}), i.e. oxygen consumption rate at Standard Temperature and Pressure (STP), O_2 is the concentration of the oxygen in the atmosphere, and V_m and K_m are equation constants. Values for V_m and K_m obtained by linear regression for the two linear regimes (2-10% and 10-20% oxygen) for showed variation due to temperature and cultivar. Respiration activity increased with temperature following the Arrhenius relationship. The Arrhenius plots at 5%, 13% and 20.8% oxygen were non-linear, which could be related to a change in the limiting enzymic reaction because each enzyme may be affected differently by change in the temperature. This shows how a practical model for designing and testing packaging can be developed from:

- a basic scientific model
- a complex system, which can be divided into linear regimes
- variations in oxygen, temperature and time, which affect reaction rates
- variations in raw materials (in this case cultivars), which affect reaction rate constants.

This work on cherries has been extended by the same authors (7) to examine the dynamics of exchange of modified atmospheres through typical packaging films. Good agreement was found between measured time concentrations of the gases and prediction equations and so these methods could be used in the design of packaging.

> *Think break*
> In the sweet cherry study, there were three cultivars – Burlat, Sunburst and Sweetheart. Burlat had the highest respiration rate; Sunburst and Sweetheart were lower and similar. This more active metabolism of the Burlat cherries made them the most perishable of the three cultivars. The differences in respiration rates decreased with the oxygen concentration. The respiration rate was drastically reduced when oxygen concentrations were below 10%.
>
> * List the packing and storage variables you would consider in developing modified-atmosphere storage for sweet cherries.
>
> * If the cherries were to be distributed in large cartons for display in supermarkets, how would you control the variables in storage and distribution?
>
> * If the cherries were in individual consumer packs, how would you select the packaging film and the storage atmosphere?

5.2.3 Water activity

Water is one of the reactants in much of food processing, and is also a significant part of all food materials. Often it is present in excess and therefore its effective concentration remains so close to constant that the relatively small changes in water concentrations encountered in food processing make no measurable difference to rates as the reaction progresses. Where water is not in excess, for example as in dried and intermediate-moisture foods, this may not be the case and the actual concentration of the water present needs to be taken into account. Water activity (a_w) is often used as a measure of the water concentration, being numerically equal to the (fractional) equilibrium relative humidity of the water over the product. A familiar example of its importance is with microorganisms, where water activity can have a major influence on growth. An instance of this is shown in Fig. 5.2, showing the effect of temperature (a sensitivity around 12%/°C) and water activity on the mould-free shelf-life of bakery products, from the work of Cauvain & Seiler (8).

Cakes can be considered as intermediate-moisture foods, with water activities between 0.65 and 0.90. At the lower levels, some yeasts can grow, and, as the water activity rises, moulds and other yeasts, and above 0.90 bacteria, also grow.

Moulds grow on the surface of cakes and therefore it is the water vapour in the surrounding atmosphere, equilibrium relative humidity (ERH), that is used as the abscissa in Fig. 5.2. The logarithm of the mould-free shelf-life was found to have a linear relationship with ERH within the range 74-90%, so this graph, or equations derived from it, can be used in shelf-life predictions. As well as controlling the water in the atmosphere surrounding the cake, the water activity can be modified by formulation changes. Such techniques can be used for intermediate-moisture foods, and many stable ambient products have been designed on the basis of understanding the effects of water activity levels on growth of possible spoilage microorganisms.

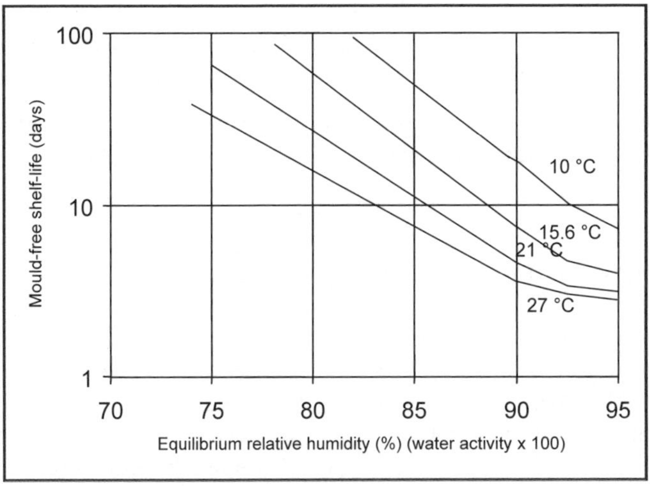

Adapted from Cauvain & Seiler (7)

Fig. 5.2. Mould-free shelf-life of cakes: effect of temperatures and humidity

Water activity also has a marked effect on chemical reactions such as non-enzymic browning and fat oxidation. In dried foods at low water activity, the fat oxidation and associated bleaching reactions are often predominant in spoilage; then they decrease in importance as the water activity increases. Browning reactions then increase, followed by growth of osmotic yeasts, then other microorganisms (9). The reaction rate effects of such water activity changes can be explored experimentally, optimum water activity ranges identified for different products, and predictive models developed.

Think break

Tomato powder, commonly used in dried soups, mixes and meals, can deteriorate on storage, losing the red colour, turning brown, developing a "hay" odour, and becoming lumpy. These changes are related to the water activity (a_w) of the powder, bleaching of the red colour and associated off-odours occurring at low a_w, and non-enzymic browning occurring at higher a_w. The bleaching is also related to the environmental oxygen.

* Discuss how you would determine the optimum a_w range where the bleaching and the browning were both at a minimum.
* How can this be related to the moisture content achieved during drying, and to the moisture content to be maintained during storage?
* Suggest how a packaging and storage system could be designed for this powder, which is marketed in bulk to manufacturers.

5.3 Alternative Energy Processing Conditions

Another approach to influencing the rates of processing reactions is by the use of higher energy processing conditions, other than thermal energy. Energy can impact on the foods either directly at a molecular level or through their environment. Both are possibilities – placing the food in intense irradiation or electrical fields, or under very high environmental pressures (*ca* 1000 atmospheres). Such systems offer the possibility of inducing and accelerating some of the various possible reactions in the food in differential patterns that are normally not the same as those in thermal reactions. By suitable selection of the energy source and level, desired reactions may be promoted, and undesired ones retarded. These physical processing possibilities all require quite elaborate equipment, more than for the conventional heating and cooling. But, because of the range of opportunities they offer, they have substantial attractions for research, development, and then industrial use.

The product outcomes, the relative qualities of the attributes of the resulting food products, can be quite different from those in heat processing. So opportunities for differentiation are offered. An important example, often the reaction critical to the processing, is that in reducing microorganisms. In particular, safety and storage life in some instances can be enhanced without any or much concomitant loss of other desirable attributes.

Such systems include the use of high-energy irradiation with particle or photon beams; high pressures (steady or intermittent); and high electric/magnetic fields (steady or pulsed). These newer processing methods, because of the new opportunities they offer, are the continuing subject of considerable investigation into their processing potential (1,10). Alternative heating methods for solid foods, where conventionally heating is by conduction, are also possible, such as microwave, radio-frequency, and resistance (ohmic) heating (11).

5.3.1 Irradiation

High-energy irradiation can be emitted from various sources and focused into the food. Sources permitted for use in processing of specified foods include gamma (γ) rays from radioactive isotopes (often Cobalt- 60), machine-generated electron beams from linear accelerators, and X-rays. The major factors determining the impact potential of the irradiation are the energy level and the flux from the source, and the time of exposure. From sources such as radioactive isotopes, the energy level is determined by the nature of the isotope – cobalt 60 has a mean energy level of 1.25 MeV; and the quantity of energy by the source strength – for example, it can be about 50,000 Curies (Ci) for fruit and vegetable disinfection and above 600,00 Ci for meat and poultry pasteurisation depending on the design of the irradiation plant. Electron beam impact is influenced by the flux of electrons and by their kinetic energies (limited by regulation to 10 MeV), and X-ray beams also by quantity and energy levels (limited to 5 MeV). The energy levels available for use are restricted to those below the potential for inducing radioactivity in the target foods, or producing unwanted chemical changes. High-energy electron beams, with their associated charge have limited penetrating power and are suitable only for foods not more than 5-10 cm thick.

The measure of the reaction-creating potential is in energy absorbed in units of Grays. The Gray (Gy) is fundamentally defined by heating capacity and measured in Joules per unit mass. One Gray is 1 Joule per kilogram, which produces very little heating; but, because of the high levels of the energy involved, the reactive impact is very much greater than that of the low-energy-level thermal radiation of the same heating power. There is a limit on the dosage; a maximum currently set by Codex Alimentarius is 10,000 Grays (10 kGy), although there could be changes in the near future.

High-energy-level irradiation in sufficient quantity destroys microorganisms. It inactivates microorganisms by damaging critical elements in the cell, such as in the genetic material. There is a loss of ability to reproduce, preventing multiplication and leading also to random termination of many cell functions (12). The extent of inhibition of growth is related to the absorbed dose in Grays by what amounts to a first order "death"/dosage relationship ("death" in this sense being assessed by absence of potential for growth). The relationship between dosage and bacterial population is not completely linear; there is an initial slower rate, a shoulder, and then a survival portion, or tail, above approximately 6 kGy. But for many practical purposes the irradiation measured in Grays can be related to a decimal reduction dose for organisms.

Therefore, it can be taken that:

$$\log (N_1 / N_0) = D_{10} / D_i$$

where D_i is the applied radiation dose, D_{10} the decimal reduction dose, N_0 the initial bacterial number, and N_1 the expected bacterial number after the irradiation.

The decimal reduction dose can be affected by the types of microorganism – spore forming and non-spore forming. The effect of the state – spores and vegetative cells – are illustrated in Table 5.II adapted from Barbosa-Cánovas *et al.* (1). Bacterial spores are more resistant to radiation than vegetative cells.

TABLE 5.II
Irradiation decimal reduction doses for microorganisms

Pathogenic bacteria	**D_{10} (kGy)**
Bacillus cereus (vegetative)	0.02 – 0.58
(spores)	1.25 – 4.0
Campylobacter	0.24 – 0.31
Clostridium botulinum	1.40 – 2.60
Escherichia coli (non-spore former)	0.23 – 0.45
Salmonella (non-spore former)	0.37 – 0.80
Spoilage bacteria	
Clostridium sporogenes	2.30 – 10.90
Micrococcus radiodurans	12.70 – 14.10
Pseudomonas putida	0.08 – 0.11
Streptococcus faecalis	0.65 – 0.70

Various sources (1,12)

An important point to note in Table 5.II is that some of the spoilage organisms require much higher reduction dosages than the pathogenic bacteria; for example, *Clostridium botulinum* requires 1.4–2.60 kGy, and *Clostridium sporogenes,* 2.30–10.90 kGy. The total range of energy susceptibilities is more compact than for thermal processing, but is still quite substantial. It is related both to the particular organisms and to whether spores can be formed, the spores being roughly one-tenth as sensitive as the vegetative cells.

Perhaps more by association from its success in canning than through any profound logic, a 12D reduction has been adopted as providing "commercial" sterilisation. So, for instance, achieving 12D values for *Clostridium botulinum* in meat systems would require dosages of 38-48 kGy according to Molins (13). Lower levels of reduction can provide pasteurisation to meet various risk-reducing, growth-inhibiting or storage-life criteria. For irradiation, there has been much consideration of the types of food and the composition in setting the standards for treatment. Indeed, in the eyes of many people, irradiation should be considered only where alternative risk-reduction or preservation alternatives are not available; it is to an extent a process of last resort.

A common target level in reduction in the United States has been 5D (5 decimal reduction) cycles (12), but for spices and dried food ingredients a 3D reduction may be adequate (14). The target and corresponding process can also be set by reduction from an initial level of bacteria in the food to an acceptable level, as shown in Example 5.2 for the reduction of *Salmonella* in chicken pieces.

Example 5.2: Irradiation to reduce *Salmonella* in chicken pieces

In a line of packed boned-out chicken pieces, there was thought to be undue risk to consumers from *Salmonella* infection. So it was decided to consider reducing initial levels, some of which had been up to 10^4 cfu/g, down to an acceptable maximum of 1 cfu in 25 g, in a Cobalt-60 irradiation facility.

It is desired to estimate the irradiation time needed, and the expected throughput, from a ^{60}Co source that can deliver 500 Gy/min to the working depth in fifty 1-kg packages, simultaneously.

The required reduction ratio is from 10^4/g to 0.04/g

that is $10^4 / 10^{-1.4}$

$= 10^{5.4}$ implying 5.4 decimal reductions

From Table 5.II, the decimal reduction dose, D_{10}, for salmonellae is 0.8 kGy = 800 Gy.

Then 5.4 D_{10} will be reached after (5.4 x 800) / 500 = 8.6 min irradiation.

So, the expected plant throughput is around (60/8.6) x 50 kg /h = 350 kg /h.

A full calculation would be more extensive than this – for example, taking into account particular geometrical factors, radiation, absorbtions, and the plant operation; but this simplified version provides some appreciation of the processing magnitudes involved.

The thickness of food is important because of the reduction of irradiation intensity on passage through a medium. Levels below the surface receive less dosage, generally receiving a logarithmic dosage reduction with distance of penetration. In Example 5.2, the dosage was estimated as being at the mean depth of penetration on a volumetric basis. So the outer layers would have substantially higher levels of reduction than the inner. For example, inactivation of a level of 10^6 cfu/g *Escherichia coli* O.147:H7 in ground beef (D_{10} =0.30 kGy) would theoretically require a minimum radiation dose of (6 x D_{10}) = 1.8 kGy as a critical limit. But if there is a 100-200% difference between the level of radiation absorbed by the food at the points of maximum and minimum exposure under commercial conditions, the ground beef could receive up to 5.4 kGy to ensure that the average dosage (on a volume basis) is adequate (15). The type of food, its composition, and the environmental conditions such as temperature alter the dosage needed for microbial safety. The moisture content, the presence or absence of oxygen, and the addition of processing agents such as nitrites in meat may also affect the dosage needed.

As with heat treatment, on irradiation there are chemical reactions occurring in the food that can cause unwanted sensory changes such as colour change in meat, fat oxidation off-flavours, and texture changes in fruit and vegetables. Sometimes, there are desirable changes such as some tenderising of meat. The reaction rates of these changes with different dosages have to be considered in the search for an optimum process.

Other irradiation possibilities are intense white or ultraviolet (UV) light. Both of these can reduce microbial loads, partly by thermal and partly by photochemical action. A disadvantage is the low penetrating power of visible and UV light, so that the most useful applications are with surface contamination such as on meats, and on thin films as with packaging materials. Systematic treatment of light processing is limited, and essentially particular cases require detailed experimentation and verification of their effects in order to give reliable process operation and outcomes.

The original research on irradiation was carried out about 50 years ago, when it was seen as a major alternative processing agent with great promise. Practical problems, some physical, some psychological, have slowed development of commercial application. The required equipment is substantial in complexity and cost, and there are major precautions needed for safety, both for personnel and for product. But in recent years, greater acceptance and in particular regulatory clearances have led to increased interest, and to commercial as well as research activity (16,17).

Think break

Irradiation is an old process, which had intensive research during the 1950s and 1960s, producing shelf-stable army rations. But it has never become a general preservation method for food products.

- Discuss what factors in food manufacturing and marketing caused this non-acceptance into the food industry.
- What new factors might cause its adoption in the next 5 years?
- Can any lessons be learned for the introduction of other new preservation methods in the future?

5.3.2 Electrical and magnetic fields

It has been observed that high-intensity electrical and magnetic fields can have major impacts on biological materials. The effects are related to the nature of the materials treated, which can include foodstuffs, packaging, and contaminants such as microorganisms. A widely investigated area is the use of pulsed electric fields (PEF), employing high-voltage gradients that have the capacity for disrupting the cellular structure and physiology, so preventing or slowing growth of microorganisms. Very high voltage gradients lead to complete breakdown of structures. But, under less severe conditions, useful reduction in pathogens and spoilage organisms can result and without, or with the minimum of the chemical or physical changes that cause quality loss in thermal processing (10).

The effect on the microorganisms relates to the strength of the electric fields, which are of the order of kilovolts/cm, the number of pulses to which the organisms are subjected, and the shape of the pulses. The sensitivity is also related to the specific organisms and to the phase of growth, being higher for cells in logarithmic growth and substantially lower for spores, as might be anticipated. Some empirical models have been suggested to describe the relation between electric fields and microbial inactivation. The simplest, assuming a first order reaction, was:

$$\ln(N_1/N_o) = -b_E (E - E_c)$$

where b_E is the regression coefficient, E is the applied electric field intensity and E_c is the critical electric field, obtained by the extrapolated value of E for 100% survival (N_1=N_o) or a survival ratio of one (E_c is related to the resistance of the particular bacteria to radiation), and N_1 and N_0 are the final and initial levels of bacterial counts. Several models have been suggested, some based on the sigmoid form of the survival relationship with strength of electric field instead of a linear relationship, the inclusion of number of pulses as well as the electric field, and a kinetic constant, k, that represents the steepness of the tangent to the sigmoid curve.

Inactivation kinetic models have also been developed that consider the effect of temperature in the first order kinetic relationship:

$$k = k_{E0}\, e^{E/RT}$$

where k is the survival fraction rate constant (μs^{-1}), k_{E0} is an electric-field-dependent variable rate constant (μs^{-1}), E is the activation energy (kJ/mol), R is the Universal Gas Constant and T is the temperature (K). This is only given as an indicative illustration, and more details can be found in Barbosa-Cánovas *et al.* (9). Inactivation kinetics for pathogenic bacteria in various types of food need to be studied, to give the basis for process safety specifications. But the general considerations illustrate how critical factors can be identified (18) and kinetic relationships identified and put to use in the new technologies.

At the present time, PEF applications are largely experimental. Large-scale commercial units would be costly and must be guaranteed safe for the operators, and more research on the treatment of different types of foods is needed to approach optimum outcomes.

Pulsed strong magnetic fields also have useful inhibitory effects on microorganisms and, like the electric fields, they offer real benefits in attacking the biochemistry and life aspects of undesirable microorganisms rather than broader chemical change. So they offer potential for processing, particularly for some classes of foods. The specifics, both of the processing and of the target susceptibilities, limit general treatment at the present state of understanding, but their potential, especially for selective effects on microorganisms, is leading to active interest in their investigation and development.

5.3.3 Very high pressures

In high pressure technology, foods are subjected to high hydrostatic pressure, generally in the range of 100-600 Megapascals (Mpa)/(1000-6000 atmospheres). Usually, this is at or near ambient temperature, although rises in temperature associated with the pressure increase can themselves add significantly to the overall outcome. Foods, in plastic film containers, are surrounded by water and high pressure compresses the water, which compresses the food. Moderately high pressures decrease the rate of growth and reproduction, and very high pressures cause inactivation of the microorganisms; spores are more resistant than vegetative cells. Under the high pressures, spores are encouraged to germinate to vegetative cells and the vegetative cells are then inactivated.

The high-pressure inactivation depends on pH, composition, osmotic pressure and temperature. Lowering the pH reduces the pressure necessary for inactivation. But temperature and pressure have an unusual interaction; generally, at moderate temperatures pressure has a synergistic effect, but at high temperatures increased pressure retards the inactivation. Water activity has a marked effect on inactivation. In one study, inactivation was greatest at 0.96 a_w, and did not occur below 0.91 a_w.

In several microbial studies, high-pressure inactivation with time followed first order kinetics, but in some cases there was a deviation because a small population of bacteria was not inactivated even after long periods of pressurisation (1). Knorr (19) noted that first order inactivation might occur, and in that case logarithmic-survivor curves can be expressed in D-values, but often deviations from this behaviour are obtained. There can be an initial lag phase indicating a certain time-dependent resistance, and also a tailing.

The thermodynamically derived equations relating the pressure to chemical reaction rates is analogous to the Arrhenius equation (18):

$$d\ln(k)/dp = -\Delta V^{++} / RT$$

The characteristic constant ΔV^{++} can take positive or negative values, producing a delayed or accelerated reaction rate with rising pressure.

Enzyme activity can be initially enhanced; and then, with increasing pressure, enzymes can be inactivated, but for some enzymes very high pressures may be necessary. Biochemical reactions are affected; for example, protein denaturation can occur with increasing pressure but it is different from heat denaturation. Because of the low temperatures, the fresh colour, flavour and aroma are retained, but textures can change – for example, softening of fruits and vegetable structures, and tenderisation of pre-rigor beef. There can be increased enzymic browning after pressurisation because of increased activity of polyphenol oxidase. Little loss of vitamin C has been reported in fruit products.

High-pressure technology was introduced as a commercial process in Japan in 1990, and later in Europe. In Japan, fruit products such jams, sauces and juices are marketed. Strawberry jam, for example, retains the fresh fruit flavour. A mixture of fruits, fruit juice and acids is placed in a plastic container and then subjected to high pressures of about 4000 atmospheres for 15 min. Pressurisation allows permeation of sugar solution into the fruits as well as preservation of the jam, which gives a shelf-life of about 17 months (1,20). An interesting development is into the use of high pressures in preparing surimi gels. Research is continuing into high-pressure technology, such as seeking quantitative assessment of the effects on critical organisms, such as on spores of *Clostridium botulinum* (21), and looking at the combined effects of heating (particularly heating which occurs during the pressurising process itself) and high pressures.

High-pressure technology has the advantages of using ambient or lower temperatures, independence of size and geometry of food, flavour, nutritional value and aroma being unaffected, uniform preservation, and reduction of processing times. The disadvantages are: the equipment is expensive and at present there is a limited throughput, often undesirable texture changes, and residual enzymes and sometimes bacteria can cause changes in storage.

Think break

New processing technologies are often expensive and cannot at present be considered as replacement for current heat processes. Therefore, their use could be in developing new products or improving old products, which could carry increased costs.

* Generate ideas for new fruit products aseptically packed, using PEF, that could be accepted in your market.
* Generate ideas for new "gel" products that could be developed using high pressure technology
* Do you think your company would consider adopting these processes – PEF or high pressure. List the reasons for not accepting and for accepting, for both processing technologies.

5.4 Combined Process Technology and the Total Process

When developing new processes and products or improving present ones, it is necessary to firstly consider the total process from the raw materials to the final consumer eating the food, and secondly identify the possible processing variables. The aim is to present to the consumer the optimum product with the product attributes that they need and want. In the past, the emphasis was often on the manufacturing process, with the distribution considered separately; the development effort was concentrated on safety, and the major processing variables, usually time and temperature, were adjusted to ensure this safety. Today, because of the needs of the consumer for "fresh", nutritionally suitable, and convenient foods, there needs to be a wider, technologically based development process. Not only the critical safety attribute is emphasised but also the important product attributes, which will not cause illness but will cause product failure. All the critical points in the total process need to be identified. Today, the Hazard Analysis and Critical Control Points (HACCP) system focuses the process variables and their effects on the rates of the critical and important reactions.

5.4.1 Hurdle technology

The additive effect of different processing variables on reactions already demonstrated in thermal operations, opens up the possibility of using different levels of several processing variables in parallel or in sequence to reach an overall product outcome. Combining different processing variables can offer significant advantages, both in control of attribute change and also in control of operations, for the efficient utilisation of equipment, production and storage. This approach has been looked at quite extensively in the literature, and given a title, *hurdle technology* (22). Hurdle technology originally emphasised safety and preservation but it can be extended to cover other product attributes.

For example, with acids in foods, pH may not be the only critical factor. The growth of bacteria, and importantly also moulds and yeasts, is influenced not only by pH but also, for many organic acids, by the fraction of the acid that is undissociated, for example with acetic acid. Equations have been derived whereby the incorporation of concentration factors such as undissociated acid as well as pH, salt and specific carbohydrates, can be used to calculate an empirical factor. If this factor exceeds a certain level then microbial spoilage should not be expected. This is substantially arbitrary, but indicates how component concentrations can be added to give a sum that indicates enhanced ability to stop or retard growth of microorganisms (23).

Hurdle technology recognises that most food preservation processes make use of more than one process variable; the levels of these process variables combine to constitute barriers or hurdles to the reactions of deterioration and decay. In most processes, one or two hurdles are applied, and then further minor or additional ones may be added to reach the required outcome (1).

As originally conceived, hurdle technology combined different preservation variables, hurdles, in order to achieve preservation effects using mild processing (24,25). These hurdles are the processing variables described throughout this book: temperature, times, concentrations of food components, natural and added processing agents, water activity, atmosphere (particularly the oxygen concentration), and, where applicable, the less usual variables such as irradiation, high pressure and pulsed electrical fields. Fifty or more different hurdles have been identified in food preservation, which can be grouped as physical, physicochemical, and microbially based (26). These variables are involved in control of raw materials/ingredients, manufacturing controls, distribution, retailing and consumer use. By combining them either at the same time or in sequence, lower levels of these processing variables can be used, and better control of product attributes achieved. It can be thought of as total processing, taking every aspect into account.

Hurdle technology is being considered for improving many current processes. For example, in UHT processing of milk, a conventional continuous flow sterilisation (CFS) process is designed to effect a 9D reduction in the concentration of thermophilic spores of normal heat resistance. However, such a process may be inadequate because of the presence of spores of greater than normal heat resistance. As has been described earlier in the book, the process is constrained at the minimal limit by the need to assure safety, and at the upper limit by the minimal changes in the sensory and nutritional attributes of the product. The milk can be made safe against these higher resistant bacteria, but the increased heat treatment will downgrade the sensory attributes. It is suggested that there is the potential use of sporicidal or sporostatic agents, such as sorbic acid, nisin, lactoferrin, phosphate and even spice and essential oils to reduce the need for such high-temperature conditions (27).

5.4.2 Sous vide

There have been various applications, for example in preserving fruits and vegetables, and in ready-to-use meals. One of the more extensive, and one that has received a good deal of industrial attention, especially in Continental Europe, goes under the French name *sous vide,* meaning literally under vacuum. This is a heat process, with temperature, time and vacuum being the three main process variables; and organic acids, pH, salt, nitrites, spices, and herbs, being secondary variables. In this, the raw materials of a food are either partially cooked, or uncooked, sealed into a plastic bag from which air is excluded so far as practicable, and then given a partial cook sufficient to give target vegetative cells something like a 6 D reduction dose. Although the extent of this cooking varies, a typical specification is for the centre to be brought to 70 °C for some minutes. Some heating specifications, quoted by Creed (28), are shown in Table 5.III.

TABLE 5.III
Some heat treatments for *sous vide* products

Treatment	Chilled life	Target organism
70 °C for 40 min	6 days	*Enterococcus faecalis*
70 °C for 100 min	21 days	
70 °C for 1000 min	42 days	
70 °C for 2 min	5 days	*Listeria monocytogenes*
80 °C for 26 min	8 days	*Clostridium botulinum* (type E)
90 °C for 4.5 min	8 days	

Adapted from Creed (28)

Thereafter, the food is chilled quickly in its package, stored and distributed for periods of up to several weeks before being finally cooked again and consumed. The distribution variables are packaging, modified atmosphere, temperature and water activity. The manufacturing and distribution specification depends on the food and the target organisms. The critical product requirement is to avoid spoilage or any danger from microorganisms. Heating temperatures are lower, and process times are longer than under conventional conditions.

The chilled life of the food depends on the raw material selection and initial preparation, timings, exclusion of air, reduced water activity, and the reduced chill temperature in distribution (below 4 °C), as well as on the partial cooking and its extent. A great deal of fine detail is indicated in the total processing specification, and can be built into computer-integrated design and manufacture. Knowledge of critical and important reactions and the effect of the levels of the processing variables on their reaction rates is essential (29). Although demanding, this detail provides a great deal more constancy of steps, including ingredients, preparations and timings, than is often customary, and so the stage is set for both optimum and very consistent product quality. It does illustrate the potential of fully and tightly organised processing, throughout the total process from raw materials to consumers. The consumers must also use the correct heating and serving conditions to achieve the designed final product quality.

The notable aspect of this form of processing is the claim to preserve, to a very considerable extent, the most desirable attributes of the fresh foods. It can be regarded as a sophisticated cook-chill process. The resulting products are claimed to compete in quality with, and be suitable for, *haute cuisine* (30), although this is not always recognised by the foodservice customers and the final consumers. Chefs would seldom agree that any processed product could reach the same level as their professional preparation and cooking, and consumers would tend to agree with them. So perhaps the most substantial outlet for these *sous vide* products can be as high-quality foodservice products, and these can be marketed for their convenience to the superior foodservice outlet preparing large numbers of quality meals. As "part" meals, they could be sold to chefs who would combine them in their own recipes. The consumer buying chilled meals in the supermarket will not want to meet the greater cost and therefore will not pay a higher price unless they are convinced that the product attributes are outstanding as compared with the more routine cook/chill products.

There have been some rather conspicuous company failures in *sous vide* products by others seeking to follow in the footsteps of the successful French products. Some causes of these failures have been:

- lack of knowledge in controlling the processing of a variety of products
- poor distribution control of temperatures and efficiency of delivery
- lack of adequately low temperatures in retail and foodservice chill cabinets
- poor understanding of consumer needs
- wrong assessment of costs and need for highly technical staff.

5.4.3 Total Process technology

The use of Total Process technology from the raw material producer to the consumer, and taking all steps into account for their effects on the product, offers the process designer considerable scope and flexibility. The whole range of available process steps is carefully assessed and advantage taken of the potential of each step to obtain new and better quality products. The final Total Process is chosen to minimise the adverse reactions but still satisfy the critical criteria taken overall. Hurdle technology and sous vide are examples of the Total Technology approach. Its proper implementation can offer major steps forward for the food industry.

Think break

The use of computer-aided design and integrated manufacturing has played an important part in the development of *sous vide* (29). They have combined several computer-friendly areas such as production planning and control, ingredients and formulation, predictive microbiology, heat transfer in foods, quality management and customer information.

* Identify which of these areas are computer-based in your company.

* For which areas are there data, but they are not in a computer form that is easy to access?

* For which areas does your company not have data?

* What other forms of data would be useful if you were planning hurdle technology in a Total Process?

* Can you see how your company could use computer-based hurdle technology in a Total Process?

5.5 Some Successes of Applied Reaction Technology

Reaction technology has been applied over the years in many aspects of food processing. In the earlier applications, the outcomes have often been reached well before the theory was worked through, only more recently can the explanations be seen more clearly. In recent years, there has been direct application of reaction technology in designing the new and improved processes. Also, the progress of knowledge, when coupled with the demonstrated practicality of the applications, has stimulated research and so led to much increased understanding and therefore improved application.

5.5.1 Canning

An area of conspicuous success, where the theoretical concepts have arguably been indispensable, has been in the broad field of microorganism destruction by the application of heat in canning, pasteurisation, and other thermal processes. The notion of first order bacterial and spore death has proved a most powerful tool, in industry, and in public health. Administered quite strictly, and with continued feedback from measurement and experimentation, its results have given great confidence to the industrial preservation of food. This has been of huge value, establishing the feasibility of safe mass production, with all the low-cost and wide dissemination of products that this brings. The very extended use of canning and the huge numbers of product items consumed each year bear eloquent testimony to the effectiveness of both the practicalities and the knowledge on which they are based. Both techniques and technology are under investigation and improvement all the time, leading to wider product ranges and higher quality. There would still appear to be considerable scope for improvements in product quality, and for extension to as-yet-undeveloped situations, particularly in heterogeneous foods and flexible packs. There have been innumerable publications on the subject, ranging from very practical guides, often put out by manufacturers of food or food equipment, to many scientific papers and books. A recent and quite comprehensive account is that by Holdsworth (31) and includes extensive tables of the kinetic factors that have been considered in this book. Heat transfer theory is used to predict temperatures in the food, and thence the rates of reactions that, integrated with time, predict product outcomes.

5.5.2 Continuous processing

Much of food manufacturing is what is called "batch" processing. This means that the raw materials are assembled, the process is started and all of the elements, subject to considerations such as good mixing, progress uniformly in time to reach the finished product. However, another possible mode is that of continuous processing; the common example is a fluid or a paste flowing through a standard pipe or an extruder. It might be heated, in which case reactions are initiated, and then cooled to terminate them. As the elements of the food move along the pipe, so the reactions occur. The product emerges at the tail of the pipe. In effect, time, sometimes termed space-time, extends along the pipe from zero at entering to total reaction time on emergence, assuming that conditions for reaction occur only in the pipe.

In continuous processing, there can be considerable economies in the space needed for a given production rate; product quality can be more uniform and control easier. But there are rigorous demands on the process understanding and instrumentation; the capital costs are often high.

Continuous processes encountered in food manufacture are largely confined to liquids passing heat exchangers, and pastes and doughs passing through extruders. A notable example is the heat treatment of milk, pasteurisation (31). Pasteurisation has extended over a whole range of liquid products, often followed by aseptic packaging. Continuous extruders are widely used to make snack foods and confectionery, and sometimes bread. Another important continuous process is the freezing of free-flow vegetables such as peas, beans, corn and small fruits.

There are also hybrids, such as continuous sterilisers for cans and pouches, and continuous freezers for fish, meat carcasses and meat in cartons. The units are in a sense batches, in cans, packets, pouches, carcasses, but they are moved through the processing region in a controlled progressing stream. As processing line throughputs increase, so continuous operation becomes more attractive.

Analysis of the process is basically similar to that in batches, if time is thought of as "space time". There are detailed treatments in reaction engineering books, but considerably simplified in foods by almost always being able to ignore heats of reaction.

> *Think break*
> Select a continuous food processing operation with which you are familiar, and a suitable book describing detail of the process (for example canning and Holdsworth (31) or milk pasteurisation and Kessler (32)).
>
> * List the ways in which residence times and temperatures, and residence time distributions, affect both the actual mean extent of processing and also any regulatory standards with which the product must comply.
> * How would you proceed to improve quality if spread of residence time distribution was found to be a major quality problem?

5.5.3 Meat freezing

Reaction technology has also provided a powerful tool to the meat industry when examining in detail apparent processing advances, particularly on the mechanical engineering of chilling and handling. With the development of more intensive refrigeration, much quicker chilling techniques became attractive in part from reduced holding times and costs, and in part from improvements in surface quality because speedier lowering of temperatures afforded less opportunity for bacteria to grow. However, feedback from customers indicated increasing dissatisfaction with tenderness of lamb carcasses that had been frozen in this way.

Physical testing of the force required to shear muscle pieces, under various cooling temperature regimes, then substantiated this toughening. The shear force required was related to the rapid lowering of temperatures. Reactions following the death of the animal were substantially slowed, as were manifestations such as slowing of *rigor mortis* and the fall in the pH, and this was found to lead to decreased tenderness. So regimes had to be devised for slower cooling with control of the maximum rate. Although the temperature/time specifications were arrived at by trial, research then quantitatively related them to the kinetics of various reactions in the *rigor mortis* development in the meat. Thus the chilling and freezing process conditions to provide acceptable tenderness in the final meat could be worked out on a systematic basis.

The knowledge base for this application was muscle physiology, and the conversion of muscle into meat. It involved understanding of *rigor mortis*, of muscular contraction, and of the breaking down of the muscle proteins and the consequent impact on the tenderness/toughness of the meat. As metabolic processes wind down, muscle moves towards the new equilibria in "meat", with changes to pH, colour, and water-holding capacity through many parallel and sequential reactions. Such reaction processes can be modified by changes in the temperatures, which have practical impacts on muscle proteins, on pH, on bacterial growth, and on meat pigmentation, juiciness, and tenderness. Some of these impacts are desirable and some undesirable.

Knowledge of the detail of some of these reactions (33) has promoted quantitative descriptions. For example, once *rigor mortis* is complete (pH has reached the equilibrium level of about 5.9), tenderising of the muscle begins at reaction rates that have an activation energy of about 60 kJ/mol. So there can be an immediate trade-off calculated between holding times before freezing, rates of tenderising and of bacterial growth, and storage temperature regimes. Other significant factors to be taken into account are water-holding capacity, colour and colour stability. As the details are becoming better known and the knowledge gaps filled, it is increasingly practicable to design processing regimes tailored to reach specific goals for meats for target markets.

5.5.4 New ingredients from milk

Milk contains a rich diversity of protein constituents, and these have been found to possess different functional properties that make them attractive, as well as nutritious, ingredients in manufactured foods. Therefore, attention has been directed to the separation of these constituents, especially from whey, which was formerly almost a waste product from cheese manufacture but now is regarded as of major value. Some of the separation can be effected by physical means, such as membrane filtration, but use is also made of the chemical properties of the different proteins, such as substantially different activation energies and rates of denaturation (33). Heating under different thermal regimes has been employed for some time, producing, for example, "high heat" and "low heat" dried milk powders, each with different functional properties and so fitted for different end uses. This has been further developed in the production of whey protein powders of differing properties, more specific in their chemical make-up and tailored more precisely to customers' requirements (34). Accounts of some of the detail can be found in Kessler (32), in Lewis & Heppell (35), and in publications of the International Dairy Federation.

5.5.5 Fresher fruit for the consumer

From observations on the rates of respiration of fruit after picking, first recorded systematically in the industrial context with apples, ideas have been generated on possibilities for control of ripening, and thus of the keeping time of the fruit. Again, temperature is a major variable, and its combination with atmospheric manipulation has led to quite dramatic extensions of the storage life of fruit. Similar combinations of packaging, atmosphere, and temperature can also be extended to vegetables.

The reaction rates follow the standard processing equations. Practical manipulation is limited by the fundamental rates of the reactions and by the economics of working at the industrial level. As with so many natural raw materials, the rates are related not only to the type of fruit but also to the individual varieties. Quite extensive exploration of the detail has been carried out for example on sweet cherries (6), in order to design processes. Once the parameters have been determined, the desired store or package atmospheres and temperatures can be prescribed. This includes prescription of the permeability of the packing material both to the outside atmosphere and to the gases formed inside the pack, as respiration and deterioration are dynamic and not static processes. Gaseous components can move across the barriers, at rates dependent on the nature of the packaging material, such as permeability and thickness and the partial pressures of the gaseous components on the two sides.

5.5.6 Food ingredients modification

For some major food ingredients, chemical reactions have long been used to modify their properties and make them more adaptable and useful. Fats and starches are major food components; a very wide range of reactions has been investigated to change their characteristics, and many have been applied industrially. The modification processes themselves are generally straightforward chemical reactions, and are described in texts such as Hui (36) for oils and fats, and van Beynum & Roels (37) for starches.

Because these start out by being novel applications in food, and their products have not been through the historical trials (and errors) of historical food selection, a major constraint must be safety in use. This is hard enough to determine for the shorter terms of days and months, let alone the lifetimes and generations that are implied in food consumption. Major problems for food products include demands on purity of ingredients, and on manufacturing practice, which must be specified and known at more demanding levels than for many other industrial processes. However, applications as in modified starches, and hydrogenated and interesterified fats, have found very widespread application because they bring such useful properties. Their manufacture lies somewhere between the chemical and the food industries, just as the technologies, though basically as has been described here, move beyond what has been briefly discussed and towards the more ambitious treatments of the chemical process industries.

Think break

Select a recently introduced novel processed ingredient (it might be a new fat, modified starch, carbohydrate gum, refined protein) and list the raw materials and the reactions required for its production. Try to find out any particular problems that were encountered in its development and introduction. Classify the major difficulties encountered as technical, legal, health, safety, commercial, consumer acceptability.

5.5.7 Storage lives

A successful and widely used food application of reaction technology has been in shelf and storage life prediction and extension. Initially, straightforward zero order, constant rate assumptions, coupled with exponential (Arrhenius or other more or less equivalent) handling of temperature effects, proved very powerful. It is useful for predicting the time that products can be held under particular temperatures and still retain necessary qualities. It is necessary for covering the conditions needed to achieve some mandated storage life, as in product life dating (38).

5.5.8 Packaging

In the context of processing, and assuming that the package is sufficient to prevent access of any organisms such as bacteria, this can be treated as an extension of controlled atmospheres to the extent that the packaging controls and modifies concentrations of potential reactants around the food itself.

The package ranges from the almost totally impermeable and inert can to plastic films of variable permeability, both generally and also to specific constituents such as reactive gases and water vapour. Gases can move in either direction, in or out of the package, increasing or decreasing their concentrations at the food interfaces depending on their partial pressure gradients. Rates of transmission depend on permeabilities and thicknesses of the packaging and on the partial pressure differential. The situation can be complicated by the reactions of the food, which can both absorb and also generate gases. These gases sometimes promote and sometimes inhibit reactions. For example, within the package, lowered oxygen partial pressure can be used to reduce oxidation, and increased carbon dioxide to reduce bacterial growth. The actual access of the gases to the reacting surfaces can be predicted from the physical situation governing concentrations at the gas/solid interfaces using the standard equations of unit operations. From these concentrations, reaction technology has been used in the usual way to predict rates and extents of the reactions. It is also used in combination with package and film selection to design packaging, to predict shelf lives, and to select conditions of storage, in particular temperature, to meet particular needs (39).

> *Think break*
> Select a packaged food produced by your organisation or familiar to you (possible choices might be a packet of cheese slices, or of fresh vegetable pieces), and list all the functions of the packaging. From these functions identify the ones that affect the rates of the reactions in the food. Find out as much detail as you can of the relationship between the package (choice of material, thickness, strength and so on) and the rates of the reactions.
>
> Focus any gaps in your knowledge that need to be explored to fully specify an optimum package for that food product.
>
> How might you find the knowledge that you need to achieve optimum storage life?

5.6 Practicalities

Looking across the broader range of processing also exposes some of the practicalities of process measurement and control that arise and which are related to the reactions in the product.

5.6.1 Quantitative product attribute measurement

So far, most of the discussion, in particular the illustrative quantitative examples, has been related to "concentrations" of product attributes, measured in physical terms, in chemical concentrations and in microbial numbers. In fact "scientific" technology is only possible if quantitative measurements can be made, its quality depending on the quality of the measurements; so in part the history of technology is the history of instrumentation. In recent times, this growth in instrumentation and its application in the food industry has been rapid, and even intractable areas such as bacterial concentrations are opening to machine measurement. However, because food is for people, there always remains one ultimate sanction, that of customer approval, and of gradations of this in assessing attributes that can be particularised in taste, such as sweetness, bitterness, and in other aspects of flavour, texture and appearance.

Part of the impetus for the developments in sensory testing arises from practicability and cost. Another aspect is more fundamental in that the sum of personal assessment of sensory attributes adds up to market judgement. Purchase choice is based strongly on personal sensory assessment of the product. So a very significant development has been the "quantification" of sensory attributes and their relation to processing and its dynamics. Sensory attribute measurement by a trained sensory panel has the accuracy needed for use in reaction kinetics (40). The correlation between sensory measurements and physical measurements has been important, which leads to simpler and cheaper quality measurement (41).

But consumer acceptability of the total product can also give quantitative measurement, as shown in Example 5.3 (42).

Example 5.3: Shelf-life of a coleslaw mix

In studying the effects of temperature on the acceptability of coleslaw, Wilkinson (42) had a panel of 20 consumers comparing acceptability of the stored samples against a reference sample, which was freshly made coleslaw. The scores used were, reference = 9, acceptable = 5 and extremely unacceptable = 1. Scores of 5-9 indicated samples that were still acceptable for the market, scores of 1-4 indicated scores that were less than acceptable. The panel tested colour, appearance, aroma, texture, flavour, and overall acceptability. The reference sample was taken as 100% acceptable and the stored sample scores were then expressed as percentages of this, to give % acceptabilities. Internal and cross testing established the consistency and reliability of their assessments so that the scale intervals behaved equally and linearly. So the panel results could be used to make relative judgements on the qualities of a range of products.

This was done with a coleslaw product with the following average results for two replicate tests, after storage at various temperatures.

Contd..

Example 5.3 (contd)

2 °C

Storage time (days)		3	7	10	16
% acceptability		95,91	82,78	71,68	54,50

6 °C

Storage time (days)		3	7	10
% acceptability		86,86	69,64	56,51

10 °C

Storage time (days)		3	7
% acceptability		80,75	50,47

21.5 °C

Storage time (days)	2	3
% acceptability	60,52	40,35

Plotting these results as shown in Fig. 5.3 displayed consistent zero order behaviour for the acceptability level as judged by the panel. Further, the Arrhenius plot gave a good straight line and a value of the activation energy of 64.7 kJ/mol and a frequency factor of 6.82 x 10 day^{-1}. This knowledge was then used in preparation, marketing and storage of the products.

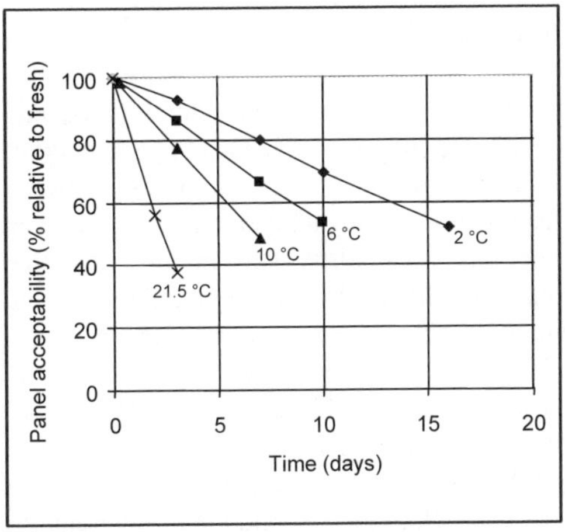

Adapted from Wilkinson (42)

Fig. 5.3. Acceptability of coleslaw as a function of time, at different temperatures

Example 5.3 demonstrates that acceptability of quality attributes and overall acceptability can be found to behave in a systematic fashion, just as do other measurable attributes such as concentrations of constituents or microorganisms, and physical measurements such as colour. On the basis of such measurements, the shelf-life to some cut-off point on the scale (one judged by the panel as representing the lowest fully acceptable quality), at actual holding temperatures can be predicted with some confidence. Alternatively, a storage programme could be planned to meet a desired storage life within the range of acceptance for this product. Or again, selection and handling of the raw materials, and the various processing steps, could be re-examined to see whether they could be re-set to give longer storage lives if this were thought to be important and feasible. Use-by dates could be supplied to retailers, printed on packets, and incorporated into the planning schedules for this short-life product.

Think break
Your local restaurant trade has asked for a line of high-quality fresh prepared vegetables, which would maintain this quality for at least one week. List the process steps you would need to introduce to set up such products from growers to restaurant. Identify the critical control points and the control procedures associated, to meet this request. What are the important reactions in the process that could affect the quality of the vegetables?

5.6.2 Temperature control

One of the practical outcomes of the reaction technology equations is that, because they show how the outcomes of processing vary with temperature, they also show quantitatively the sensitivity of these outcomes to temperature changes. Conversely, they display how closely temperature has to be controlled to maintain a particular level or band of quality. This in turn indicates the demands on temperature controllers and hence the precision and reproducibility demanded of thermometric elements (thermometers). For instance, if the reaction has a sensitivity of 25%/°C, a thermometer accurate only to 1 °C allows a possible 25% shift in quantitative outcome, and this could well be outside the specifications demanded in many circumstances. Therefore, instruments and controllers need to be considered carefully in regard to the precision of their action, and in turn related to the process technology details and the product quality required.

5.6.3 Measurement of process extent

In quite a number of industrial processing situations, it is difficult, and sometimes impossible, to make measurements on the primary critical attributes as they change during the processing to the degree of precision required for proper control. For example, in canning sterilisation, where there is a nominal 10^{12} ratio between the initial and final bacterial spore counts, and the final count therefore is equivalent to 1 viable measurable spore in perhaps 10^9 cans, there is no practical way by which this end condition can be directly measured. Imagine conducting spore counts on a thousand cans, let alone a thousand million. The outcome of a sterilising process can be inferred from temperature records through calculations, as has been seen. But it would be very helpful if there were some accessible single measurement that could be taken and which would have direct and unequivocal correspondence to the final bacterial state but without having to take bacteriological measurements or estimates.

As has also been seen, there are chemical changes in the product, which proceed in parallel with the bacterial sterilisation and which may be measurable. But if such inherent reactions are not convenient, then selected cans could be "spiked" with chemicals, which would "piggy-back" with the critical bacterial reaction and would themselves change in a well-behaved manner that can readily be measured. There have been extensive published discussions of these problems, and reviews (43).

The fundamental problem is: what correspondence is there between such a chemical reaction and the bacterial death; indeed, between any two parallel systems of different reactions constrained to follow identical temperature histories?

Reflection on the kinetics (Theory 5.1) shows that there will be direct correspondence if, and only if, the activation energies of the parallel reactions are identical. If they are different, then the degree of correspondence will depend on the temperature path and therefore can only be known if the temperature path itself is also known.

Theory 5.1: Integration of parallel reactions – analogues

Taking two first-order reactions:

$$dC_A / f(C)_A = - k_A \, dt \qquad dC_B / f(C_B) = - k_B \, dt$$

and so, on dividing, $\{dC_A / f(C)_A\}/\{dC_B / f(C)_B\} = - k_A / k_B$

The LHS includes only initial and terminal compositions C_{A0}, C_A, C_{B0}, C_B and is independent of temperature if k_A and k_B are both constant, or if their ratio is constant and independent of temperature.

Since from the Arrhenius equation,

$$k_A/k_B = A_A/A_B \exp\{(E_B - E_A)/RT\},$$

this can only be independent of T,

$$\text{if } (E_B - E_A) = 0,$$

that is, if the activation energies of the two reactions are equal.

In fact, the potential for altering relative process outcomes in different components by manipulation of temperatures during processing depends on such variation. If all temperature paths led to identical relative outcomes for all constituents, then choosing different temperature paths would not affect the outcomes. That this is not true is shown from industry experience, generally, in Theory 5.1, and in specific instances by Example 5.4 and by Fig. 4.7.

These problems can perhaps best be illustrated by a simple example, using a process with two constant temperature steps, in Example 5.4. More complicated processes will compound the problems.

Example 5.4: Parallel reactions: sterilisation and thermal inactivation of trypsin

Thermal inactivation of trypsin, a first order process, has been advocated as a model for first-order, high-temperature, thermal processes for sterilisation (destruction of 10^{12} thermophilic spores), where full experimental measurements covering the thermophilic spores are not possible. A suggestion has been to use the inactivation of the enzyme trypsin as an analogue to be followed, with the idea that this can be related to the sterilisation process.

If the equation for the rate constant of inactivation of trypsin is:

$$k = 8 \times 10^{9} \exp(- 1.01 \times 10^{9}/T) \text{ min}^{-1}$$

and for the sterilisation of typical thermophilic spores is:

$$k = 1.2 \times 10^{42} \exp(-3.73 \times 10^{4} /T) \text{ min}^{-1}$$

Determine the total changes (∇_{spores}, $\nabla_{trypsin}$) for two heat processes, each containing two constant temperature steps in which:

(a) 2.0 min at 116 °C is followed by 1.5 min at 123 °C
(b) 3.6 min at 116 °C is followed by 0.5 min at 123 °C

From the rate constant equations:

for spores, $k_{116} = 2.73$ min^{-1} and $k_{123} = 14.86$ min^{-1}
for trypsin, $k_{116} = 0.042$ min^{-1} and $k_{123} = 0.067$ min^{-1}

and so for (a) $\nabla_{spores} = (2.0 \times 2.73) + (1.5 \times 14.86) = 27.8$ (= 12.1 D)
$\nabla_{trypsin} = (2.0 \times 0.042) + (1.5 \times 0.067) = 0.185$ (= 0.08 D)
and for (b) $\nabla_{spores} = (3.6 \times 2.73) + (0.5 \times 14.86) = 17.3$ (= 7.5 D)
$\nabla_{trypsin} = (3.6 \times 0.042) + (0.5 \times 0.067) = 0.185$ (= 0.08 D)

Contd..

Example 5.4 (contd)

These results demonstrate clearly the problems of trying to predict from the integrated trypsin inhibition of 0.08D in each case, the integrated results for the spores, 12D in one case and less than 8D in the other. Neither of the processes exemplified was at all extreme. Choosing more extreme cases will lead to more dramatic discrepancies.

Because of the practical importance of being able to conveniently monitor industrial reactions in foods, a great deal of ingenuity and numerous patents for analogue systems have emerged over the years. Chemical, electrical, and physical systems, including enzymic reactions, electrodeposition, and diffusion, with measurements by weighing, densitometers, colour changes, voltages, currents and spectral absorption, have all been advocated. The critical issue, apart from reproducibility and adequate accuracy, remains that of the activation energies, and, if in doubt about any particular application, then it is always wise to conduct your own checks. These are quite simple, although those for product storage may take quite a time.

Think break

Continuous canning presents great difficulties for recording temperatures through the process and therefore predicting the destruction of *Clostridium botulinum* in the cans. Suggest how you could seek an analogue that would enable the prediction of the extent of sterilisation accomplished in continuous canning.

5.7 Opportunities for the Future

Reaction technology applications belong not only to the past and the present, but they also offer prospects for the future growth and development of the food industry as craft skills are increasingly supplemented by systematic application of science and scientific methods, and as understanding of these methods grows.

5.7.1 More uniform product quality

Greater knowledge of the effect of process variables on reaction rates, and more exact measurements and control can lead directly to more precise product outcomes. All parts of the food should receive just that level of processing that is appropriate. This may mean uniformity, but on the other hand it may mean levels that are differentiated for different parts of the food, for instance, the crust on bakery products or on roasted meats.

5.7.2 Nutritional enhancement

Very often, nutritional value decreases on processing, as labile components such as vitamins may be destroyed in chemical reactions whose rates are increased by higher temperatures. Proteins may be denatured or broken down to peptides with diminished nutritional value. On the other hand, access to nutritionally beneficial components may be improved by heating, such as denaturation of trypsin inhibitors in beans or reduction in cyanides in cassava. In any event, knowledge of the basic components and of the reactions they can undergo can be used to design processes that will benefit consumers' nutrition.

5.7.3 Safety

Freedom from the effects of harmful components, and in particular microorganisms that can produce toxins or are pathogenic, can be ensured only by adequate knowledge of their presence and concentrations, and then the application of the necessary and controlled processing needed to reduce these to levels that are safe. To ensure this means continued improvement of processing technology, both to deal with present and known hazards and to be equipped to meet the new ones that will surely emerge, or grow in significance, in the future.

5.7.4 Better and more effective regulation

Greater understanding, and more comprehensive appreciation of the scope, reliability and power of modern simulation and prediction techniques must lead in time to more harmonious and rational conceiving, drafting and implementation of regulatory regimes. It will provide both the regulatory authorities and the industry with a much more substantial base for making decisions that are important and often far-reaching. This is already being seen in many jurisdictions. With the scientific basis and justification of the technology more clearly demonstrable, this should allow for more flexible and adaptable processing to give simultaneously safer, more nutritious, and more desirable foods, making the best use of the ingredients and processing possibilities.

5.7.5 Technological skills

Reaction-based process technology offers a comprehensive and rational basis for understanding the changes that the food industry makes to its products during their manufacture. This can be taught, and will add to staff knowledge, adaptability, confidence, and creativity. It lifts the level of understanding, moving from a craft towards a profession while still retaining and enhancing the skills to produce quality foods. It offers an attractive path to the recruits of the future, smaller in number but more advanced in education, and capable of dealing with the information revolution that the food industry, like all industries, can turn very much to its advantage through superior technology.

5.7.6 Instrumentation and automation

Processing can only be applied under full control if the changes made in product attributes can be measured, and their extent can be determined and ensured. So adequate measuring instruments, with precision and accuracy equal to or at least commensurate with the quality levels demanded by the consumer, must be available. These can be operated manually. But increasingly and desirably, this should be through automatic control equipment, regulating the process, batch by batch, line by line, item by item for today's, tomorrow's, and future production. The better the knowledge of the process, the more completely all of this can be put in place.

5.7.7 Optimisation

Getting the best value for least resource expenditure is always important. So, although trial and error can often provide good solutions, those errors can be costly and time-consuming. Perhaps more importantly, situations are continually shifting and with them the optima, and so the facility to adapt to change is central to good operation. Reaction technology from its very nature can deal with changing numbers, such as concentrations and time, and so, as technology is improved, so attainment of overall optimisation comes closer and closer.

5.7.8 An enhanced basis for food reaction technology

This book has used simplified structures and techniques for the quantitative description of the processing. More elaborate understanding and correlations already exist for parts of food processing – for example, Van Boekel and Tijskens (44) – but they have not been introduced here in part to maintain simplicity, but also because so often neither the available data nor the quality demands of the product justify added complexity. However, in the future, more extensive use of dedicated computer software, increased accumulation of practical data, experience and confidence, and better understanding of the theory will no doubt happen. Also, as the understanding of the underlying chemical changes and their mechanisms extends, this can be used to make predictions of behaviour and so of what might be industrially feasible. So there will be much extended scope for true designer foods and food processes.

5.7.9 New food products

Technology as it opens up possibilities for new, and sometimes unexpected, change can offer access to novel products. This can include both totally new products, such as the whey protein isolates, new products from new processes or new twists on old processes such as the *sous vide* lines, and improvements on old products or ingredients such as the pasteurised liquid whole egg. As both techniques and understanding improve and extend, so new possibilities will appear and new products will reach the market. Increasingly, it will be possible to design these products and production lines systematically, and the technology of the reactions as they are involved in industrial manufacture will be an essential quantitative element in such design.

5.8 References

1. Barbosa-Cánovas G.V., Pothakamury U.R., Palou E., Swanson B.G. *Nonthermal Preservation of Foods.* New York, Marcel Dekker, 1997.

2. Chen C-C., Tu Y.-Y., Chang H.-M. Thermal stability of bovine milk immunoglobulin G(IgG) and the effect of added thermal protectants on the stability. *Journal of Food Science*, 2000, 65 (2), 188-93.

3. Tijskens L.M.M., Biekman E.S.A. pH in action, in *Proceedings of the International Symposium on Applications of Modelling as an Innovative Technology in the Agri-Food-Chain.* Edited by Hertog M.L.A.T.M., Acta Horticulturae, 566, 2001.

4. Ryan-Stoneham T., Tong C.H. Degradation kinetics of chlorophyll in peas as a function of pH. *Journal of Food Science,* 2000, 65 (8), 1296-302.

5. Eburne R.C., Prentice G. Modified-atmosphere-packed ready-to-cook and ready-to eat meat products, in *Shelf Life Evaluation of Foods*. Edited by Man C.M.D., Jones A.A. London, Blackie, 1994.

6. Jaime P., Salvador R.O., Oria R. Respiration rate of sweet cherries: 'Burlat', 'Sunburst', and 'Sweetheart' cultivars. *Journal of Food Science,* 2001, 66 (1), 43-7.

7. Salvador M.L., Jaime P., Oria R. Modeling of O_2 and CO_2 exchange dynamics in modified atmosphere packaging of 'Burlat' cherries. *Journal of Food Science,* 2002, 67 (1), 231-5.

8. Cauvain S.P., Seiler D.A. *Equilibrium Relative Humidity and the Shelf Life of Cakes.* Report 150, Flour Milling and Baking Research Association, Chorleywood, 1992.

9. Jones H.P. Ambient packaged cakes, in *Shelf Life Evaluation of Foods, 2nd Edn.* Edited by Man C.M.D., Jones A.A. Gaithersburg, MD, Aspen, 2000.

10. Barbosa-Cánovas G.V., Góngor-Nieto M.M., Pothakamury U.R., Swanson B.G. *Preservation of Foods with Pulsed Electric Fields.* San Diego CA, Academic Press, 1999.

11. Ohlsson T. Minimal processing of foods with electric heating methods, in *Processing Foods: Quality Optimisation and Process Assessment.* Edited by Oliveira F.A.R., Oliveira J.C. Boca Raton, CRC Press, 1999.

12. Dickson J.S. Radiation inactivation of microorganisms, in *Food Irradiation: Principles and Applications.* Edited by Molins R.A. New York, John Wiley & Sons, 2001.

13. Molins R.A. Irradiation of meats and poultry, in *Food Irradiation: Principles and Applications.* Edited by Molins R.A. New York, John Wiley & Sons, 2001.

14. Farkas J. Radiation decontamination of spices, herbs, condiments and other ingredients, in *Food Irradiation: Principles and Applications.* Edited by Molins R.A. New York, John Wiley & Sons, 2001.

15. Molins R.A., Motarjemi Y., Kaferstein F.K. Irradiation: a critical control point in ensuring the microbiological safety of raw foods, in *Irradiation for Food Safety and Quality*. Edited by Loaharanu P., Thomas P. Lancaster, PA, Technomic, 2001.

16. Molins R.A. (Ed) *Food Irradiation: Principles and Applications.* New York, John Wiley & Sons, 2001.

17. Loaharanau P., Thomas P. *Irradiation and Food Safety and Quality*. Lancaster, PA, Technomic, 2001.

18. Wouters P.C., Alvarez I., Raso J. Critical factors determining inactivation kinetics by pulsed electric field food processing. *Trends in Food Science and Technology,* 2001, 12, 112-21.

19. Knorr D. Process assessment of high-pressure processing of foods: an overview, in *Processing Foods: Quality Optimisation and Process Assessment.* Edited by Oliveira F.A.R., Oliveira J.C. Boca Raton, CRC Press, 1999.

20. Wan X., Melton L.D. High-pressure preservation of food. *New Zealand Food Journal*, 1999, 29 (6), 220-2.

21. Sizer C.E., Balasubramaniam V.M., Ting E. Validating high-pressure processes for low-acid foods. *Food Technology,* 2002, 56 (2), 36-42.

22. Leistner S., Rödel W. The stability of intermediate moisture foods with respect to microorganisms, in *Intermediate Moisture Foods.* Edited by Davies R., Bench G.G., Parker K J. London, Applied Science, 1976.

23. Jones A.A. Ambient-stable sauces and pickles, in *Shelf-Life Evaluation of Foods, 2nd Edn.* Edited by Man C.M.D., Jones A.A. Gaithersburg, MD, Aspen, 2000.

24. Gorris L.G.M., Tauscher B. Quality and safety aspects of minimal processing technologies, in *Processing Foods: Quality Optimisation and Process Assessment.* Edited by Oliveira F.A.R., Oliveira J.C. Boca Raton, FL, CRC Press, 1999.

25. Leistner L. Minimally processed, ready-to-eat and ambient-stable meat products, in *Shelf-Life Evaluation of Foods.* Edited by Man C.M.D., Jones A.A. Gaithersburg, MD, Aspen, 2000.

26. Ghazala S., Trenholm R. Hurdle and HACCP concepts in *sous vide* and cook-chill products, in *Sous Vide and Cook-Chill Processing for the Food Industry.* Edited by Ghazala S. Gaithersburg, MD, Aspen, 1998.

27. Reid D.H., McCarthy, O.J. Development in the aseptic processing of milk and other foods: a critical review. *NZ Food Journal,* 2000, 30 (3), 98-107.

28. Creed P.G. Sensory and nutritional aspects of *sous vide* processed foods, in *Sous Vide and Cook-Chill Processing for the Food Industry.* Edited by Ghazala S. Gaithersberg, MD, Aspen, 1998.

29. Martens T., Nicolaï B. Computer-integrated manufacture of *sous vide* products: the ALMA case study, in *Sous Vide and Cook-Chill Processing for the Food Industry.* Edited by Ghazala S. Gaithersburg, MD, Aspen, 1998.

30. Ghazala S. *Sous Vide and Cook-Chill Processing for the Food Industry.* Edited by Ghazala S. Gaithersburg, MD, Aspen, 1998.

31. Holdsworth S.D. *Thermal Processing of Packaged Foods.* London, Blackie Academic, 1997.

32. Kessler H.G. *Food Engineering and Dairy Technology.* Freising, Verlag A. Kessler, 1981.

33. Young O.A., Lovatt S.J., Simmons N.J. Carcass processing: quality controls, in *Meat Science and Applications*. Edited by Hui Y.H., Nip W-K., Rogers R.W., Young O.A. New York, Marcel Dekker, 2001.

34. Huffman L.M. Processing whey protein for use as a food ingredient. *Food Technology*, 50 (2), 49-53, 1996.

35. Lewis M.J., Heppell N.J. *Continuous Thermal Processing of Foods*. Gaithersberg, MD, Aspen, 2000.

36. Hui Y.H. (Ed.) *Bailey's Industrial Oil and Fat Products. 5thEd.* New York, Wiley, Interscience, 1996.

37. van Beynum G.M.A., Roels J.A. (Eds) *Starch Conversion Technology.* New York, Dekker, 1985.

38. Man C.M.D., Jones A.A. (Eds) *Shelf-Life Evaluation of Foods. 2nd Edn.* Gaithersburg, MD, Aspen, 2000.

39. Brody A.L. Intelligent packaging improves chilled food distribution. *Food Technology*, 2001, 55 (10), 85-6.

40. Meilgaard M., Civille B.S., Carr B.T. *Sensory Evaluation Techniques.* 3rd Edition. Boca Raton, FL, CRC Press, 1999.

41. Munoz A.M. *Relating Consumer Acceptance and Laboratory Data.* (*PCN28-030097-36*). West Conshohocken, PA, American Society for Testing Materials, 1997.

42. Wilkinson B.H.P. *Extended Shelf Life of Vacuum Packed Chilled Coleslaw and Salad Products*. Internal Report, Food Technology Research Centre, Massey University, 1982.

43. Hendrickx M., Maesmans G., De Cordt S., Noronha J., Van Loey A., Tobback P. Evaluation of the integrated time-temperature effect in thermal processing of foods. *Critical Reviews of Food Science and Nutrition*, 1995, 35 (3), 231-62.

44. Van Boekel, M.A.J.S., Tijskens L.M.M. Kinetic modelling, in *Food Process Modelling.* Edited by Tijskens L.M.M., Hertog M.L.A.T.M., Nicolai B. Cambridge, Woodhead, 2001.

INDEX